harpooned

Frontispiece Japanese whalecatcher in the Antarctic.

By courtesy of Nippon Suisan Kaisha Ltd, Tokyo

THE STORY OF WHALING

harpooned

by Bill Spence

CONWAY MARITIME PRESS
GREENWICH

TO
JOAN
WITHOUT WHOSE HELP AND ENCOURAGEMENT
THIS BOOK WOULD NOT HAVE BEEN WRITTEN
AND
TO
ANNE, GERALDINE, JUDITH AND DUNCAN

First published in Great Britain in 1980 by
Conway Maritime Press Ltd,
2 Nelson Road,
Greenwich, London, SE10 9JB

ISBN 0 85177 167 X

Designed by Jon Blackmore
Printed and bound in Great Britain by
Butler & Tanner Ltd, Frome
Phototypeset by Trident Graphic Ltd,
Reigate, Surrey

Contents

Acknowledgements

In the writing of this book I have received assistance from many people; this has come in the form of a letter or a word, of lengthy correspondence or the search for photographs, of advice and criticism. The list of people who have helped me would be too long to quote in full here; they come from many countries, and to all of them I say a very grateful thank you. I must however, make special mention of the staff of York City Library, who have been untiring in searching out books and documents for me. My sincere thanks go to the Wardens and Brethren of the Corporation of the Hull Trinity House, to the authorities of the Hull Museums and of the Hull City Library, and to the authorities of the Whitby Literary and Philosophical Society, for permission and facilities to photograph their whaling exhibits, and to Mr and Mrs G Wood of Oswaldkirk who helped in this field; I also thank all those individuals who have given or supplied photographs. I am grateful to Gerald Pollinger and my publisher for the help and guidance they have given me, and to my typist who undertook the laborious task of typing the manuscript. My very special thanks go to my wife Joan, who has done a tremendous amount of unseen work, reading, researching, filing and advising. She has dealt with all the correspondence which accumulates in writing a book of this nature, and her assistance, together with that of our family, has proved invaluable. This book would not have been written without such devotion and encouragement. Finally, my thanks go to all those whalemen who ever sailed the oceans and braved the dangers they encountered in hunting the whale – without them, there would have been no history to relate. My sympathy goes to the whale. It is hoped that this mammal will survive the prodigious slaughter which has gone on, not only in recent years, but throughout the centuries, so that it will continue to fascinate and intrigue mankind in forthcoming generations.

Bill Spence

1 Early Whaling

The light breeze gentled the mist, lying low along the North Sea coast, into a weaving, twisting pattern. It moved and swirled around a group of people, clad in skins and armed with primitive weapons, who searched the shore for food.

The leader's ponderous steps carried him ahead of the others who were strung out along the sand. He hated shore fog which was cold and clinging but today he was hopeful that the breeze would soon clear it away. He glanced up from his search, his eyes probing the thinning mist.

Suddenly he stopped. His eyes, widening with fear of the unknown, were tinged with curiosity and suspicion. The clearing fog was slowly revealing a darker mass almost as high as himself and certainly much wider: a barrier across his path.

The breeze stiffened, drawing the veil aside. The man edged forward slowly, needing to satisfy his curiosity and not to appear afraid to the group he led. The mass took shape. A creature lay still upon the sand. A shiny, black body tapered from a round head to a tail. It was long, about four times as long as the man was tall and proportionately more massive.

He stared in wonderment at the huge creature and, as he edged forward carefully and circled to his right, he saw the black skin give way to white on the underside. He stopped, amazed that such a big thing should come from the sea as this had obviously done.

He knew fish. He had caught them in the rock pools and shallows. They were good to eat. But they were small. He had not seen one like this. Maybe this was good to eat. If it was then – so much food! Sufficient to feed his band for many a day!

Driven by the possibilities he stepped towards the animal, keeping his eyes fixed on the head, ready to retreat at the first sign of movement. But there was none. He reached out gingerly with the long stick he carried. The stick made contact. He prodded and then, anticipating a reaction, jumped back. He gazed at the form. It still lay prone. There was no movement at all.

This Neolithic rock carving from Skegerveien, Drammen, Buskerud, Norway shows that Stone Age Man was familiar with the whale, no doubt from those cast up on the shore.

Universitetet Oldsaksmaling, Oslo, Norway

Feeling easier, a little more daring, the man stepped closer. He pushed his stick against the smooth skin and prodded harder. He repeated his action several times, quickening his thrusts. With each one came a lessening of the tension and apprehension which he had felt when he first sighted this unknown creature. This thing was not going to rebel against his prodding. It was dead.

Satisfied, the man threw back his head and sent out a call to penetrate the last vestiges of the fog.

The answering calls mingled with the thud of feet running across the sand. Men, women and children joined him and all were as astonished as he at their first sight of this strange creature which, even in death, had entered their world.

An order came from the leader and every adult fell upon the body with their stone weapons while the children ran around the animal in playful curiosity. Men and women hacked at the skin until they had cut a way through to the thick, fleshy substance below it. All eyes turned to their leader. He cut a thin slice. Everyone watched him hopefully. They were hungry. They wanted food.

He stared at the oily softness in his hand. A quick lick with his tongue, a savouring of the taste, a grunt and then he bit a piece from the slice. The doubt disappeared from his face and with a grin he grunted his approval. The tension went out of his followers; shouts of joy rose on the sea air and they turned back to the body to cut eagerly at the flesh. Tonight they would sleep with full bellies.

A scene from the imagination as it flies back through the centuries to primitive man, but one which could have happened not only along the shores of the North Sea but also along other shores throughout the world. Man had discovered the whale.

Indications that primitive man used the carcases of stranded whales have been found in his refuse piles around the North Sea. The whale is depicted in Stone Age rock carvings in Norway, but there is no evidence that it was hunted at that period.

With the development of the boat, it is likely that man's hunting instincts and his knowledge of the usefulness of the whale would prompt him to pursue it close to the shore. Now he had no need to rely on stranded whales – now he could hunt it in its own habitat. From that moment the whale was doomed, for, throughout the centuries to the present day, the whale would be hunted and slaughtered to the point of extinction. The hunting would wax and wane, would be important or insignificant, but ever the hunt would go on until even today there are some nations which carry on the ruthless slaughter, taking no notice of those who warn that, if the killing does not stop, a fascinating, harmless creature, which can boast the biggest animal ever to inhabit the earth among its species, will disappear for ever.

Man's inventiveness has brought his basic weapons and implements to a stage where no whale is safe from the hunter. Way back in the mists of time those early hunters, whose basic weapon was the spear, unknowingly laid the foundation for the greatest slaughter of animals ever to take place on this planet.

Whales have always been abundant along the Pacific coast of America and there is little doubt that the whaling methods of the American Indians developed over the centuries from those of the earliest inhabitants of this coast. Once the Indian started pursuing the whale he continued without interruption until the coming of the white man. The Eskimo, too, until contact was made by white whalemen, lived in isolation parallel to that of Stone Age man. As whaling in most parts of the world developed along similar lines, it is possible to obtain some idea of how whales were hunted in early times by a study of the Indian and Eskimo methods, as recorded at a much later period by eyewitness accounts.

In 1874 Charles M Scammon wrote: '... The animal is approached near enough to throw the harpoon when all shout at the top of their voices. This is said to have the effect of checking the animal's way through the water, thus giving an opportunity to plant the spear in its body, with lines and buoys attached. The chase continues in this wise until a number of weapons are firmly fixed ... As soon as the animal becomes much exhausted ... he cuts a hole in its side sufficiently large to admit the knife and mast to which it is attached; then follows a course of cutting and piercing till death ensues, after which the treasure is towed to the beach in front of their huts, where it is divided ...'

Scammon was writing about the method of hunting whales by the Eskimos of his day but he could well have been describing the hunt as carried out by early man, who found the whale an aid to survival and a source of food, clothing and building materials for his house.

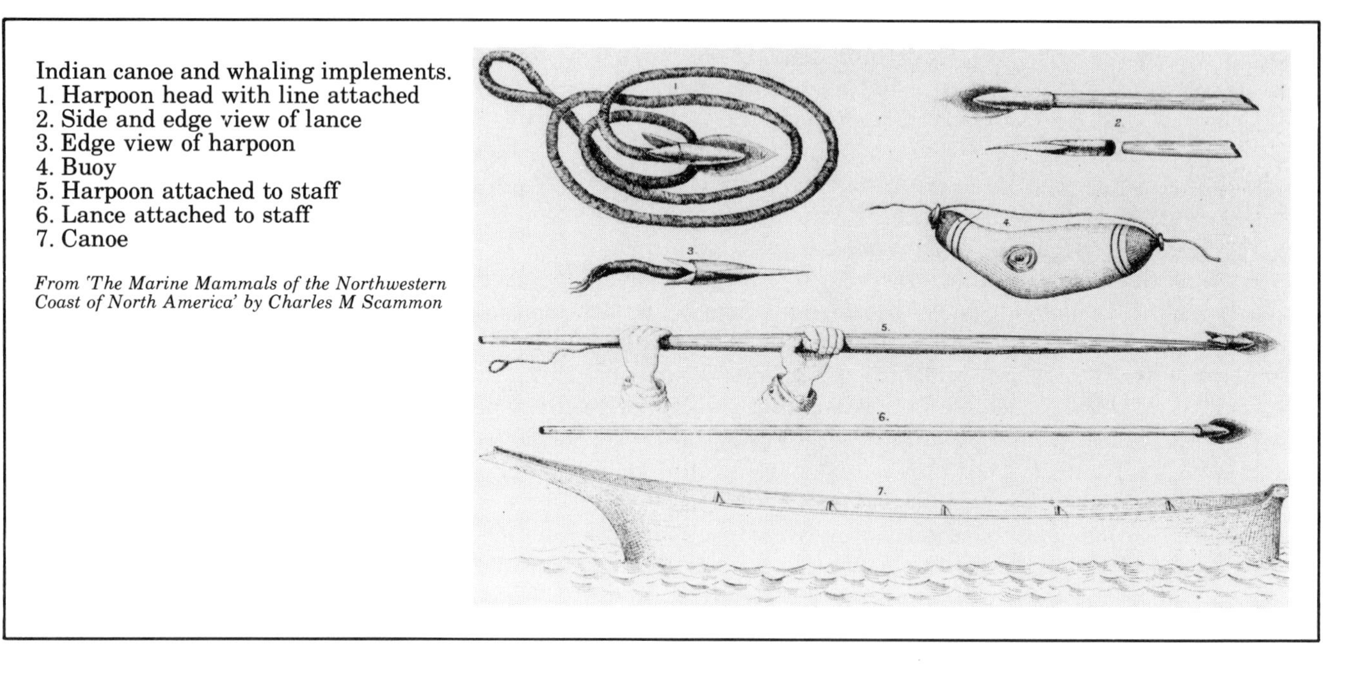

Indian canoe and whaling implements.
1. Harpoon head with line attached
2. Side and edge view of lance
3. Edge view of harpoon
4. Buoy
5. Harpoon attached to staff
6. Lance attached to staff
7. Canoe

From 'The Marine Mammals of the Northwestern Coast of North America' by Charles M Scammon

Later when contact between various communities was made, lines of communication were opened up and led to the development of trade. To the people of coastal regions, the whale soon became an important trading proposition, so much so that early maritime nations, quick to see the possibilities, opened up the whaling trade.

The Phoenicians left little writing about themselves but we know they were great seamen and traders. They would know of the sperm whale which frequented the Mediterranean at that time and it seems likely that they hunted it. If so then these expert traders would make use of it in their commercial enterprises.

The Cretans hunted dolphins in the Aegean some 1000 years after the Phoenicians. The dolphin was held sacred in many areas and appears on seals, pottery and other artefacts of the Cretan and Mycenaean civilisations, and as early as 2000BC appears in the colourful wallpaintings of the Queen's Megaron in the palace of Knossos in Crete.

References to the whale appear in the writings of Aristotle (384–323BC) and of Pliny the Elder (AD23–79), but neither the Greeks nor the Romans seem to have had any desire to hunt it.

It was left to the people living along Europe's western seaboard to develop their whaling into commercial enterprises from which the great whale trade sprang.

In his writings King Alfred the Great (871–899) refers to an account given to him by a Norseman, Othere, which reveals that the Norsemen were whaling off the coast of Norway. They practised a special method which, no doubt, had been used by man over a long period to catch the smaller whales. The doomed animals were driven into small fjords and bays which became so crowded that many of them died and, with the only means of escape blocked by the hunters, the remainder were trapped and killed. A similar method was used by ninth century Icelanders and in modern times it has been witnessed in the Orkneys, the Shetlands, Newfoundland and the Faeroes.

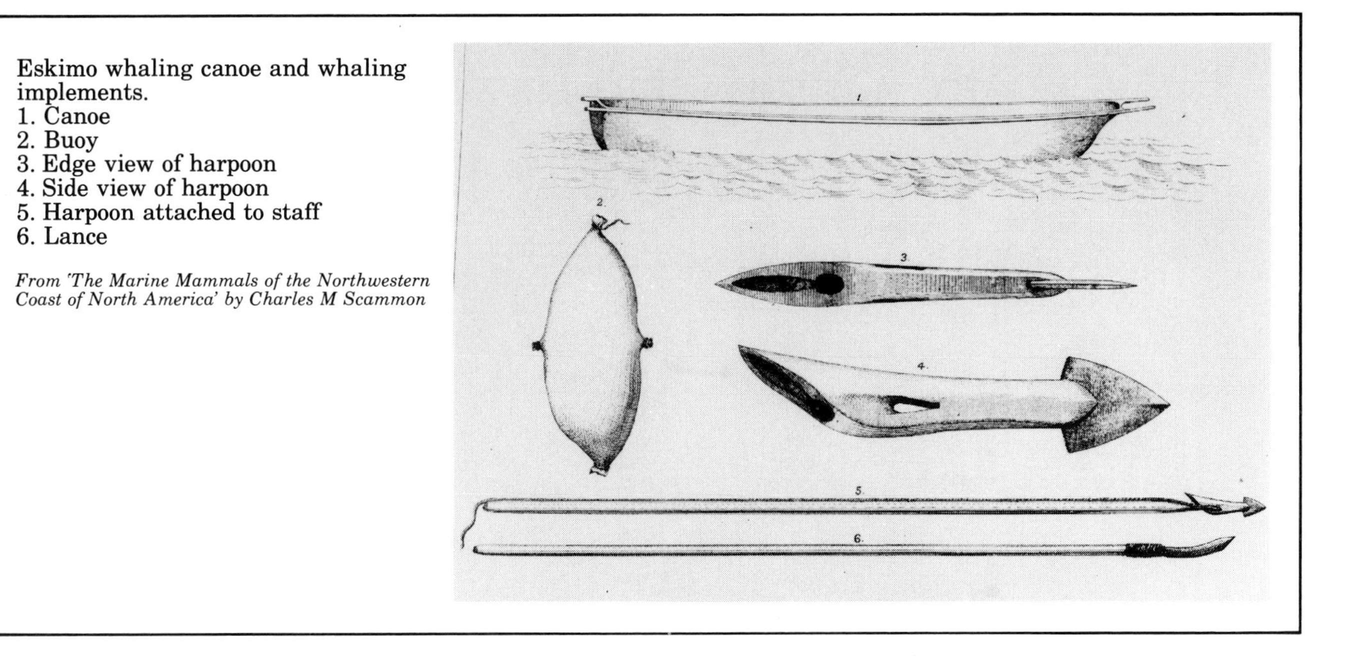

Eskimo whaling canoe and whaling implements.
1. Canoe
2. Buoy
3. Edge view of harpoon
4. Side view of harpoon
5. Harpoon attached to staff
6. Lance

From 'The Marine Mammals of the Northwestern Coast of North America' by Charles M Scammon

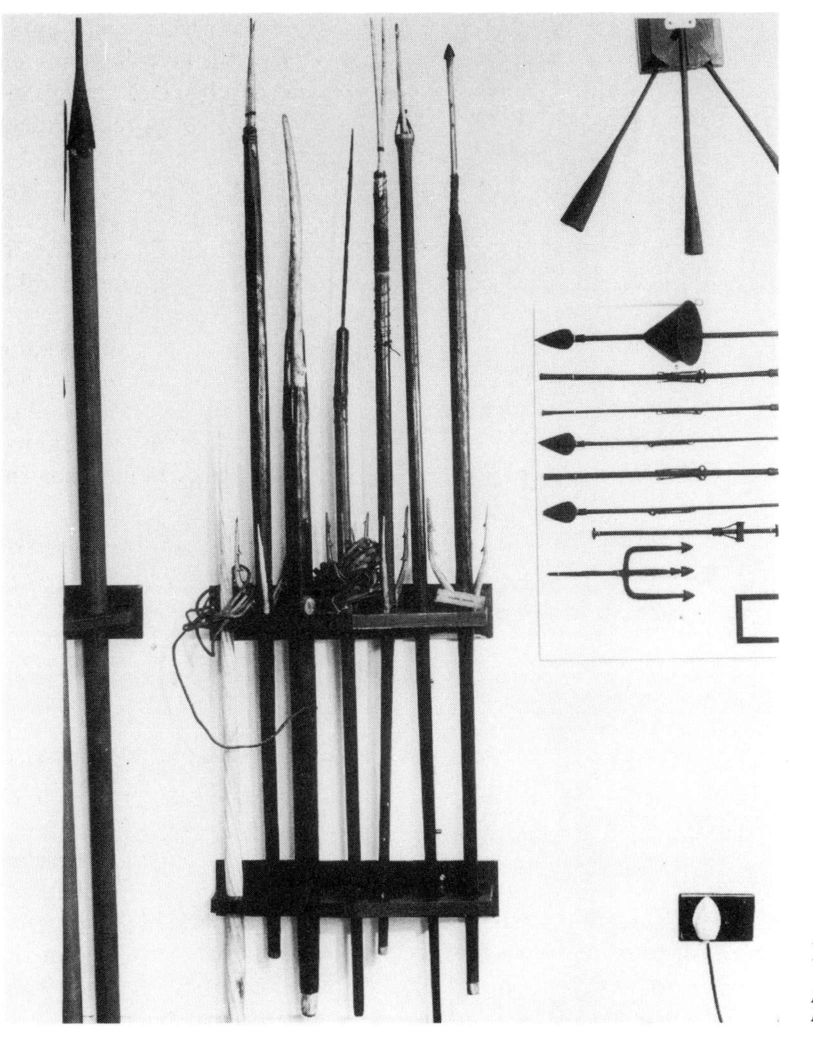

Eskimo harpoons.

By courtesy of the Whitby Literary and Philosophical Society

The activity of four men cutting up a whale is used to illuminate the letter H in 'Jonsbok', a fourteenth century Icelandic manuscript.

Arnamagnaeanske Institute, Copenhagen

There is mention of a whale fishery along the French coast in AD875, and just over a hundred years later the Viking colonists in Greenland were finding the whale useful in facing the harsh winters. Medieval woodcuts depict the whale most often as a fearsome creature. In the twelfth and thirteenth centuries it is featured in a number of gifts and dues.

By acts passed in 1315 and 1324 all whales cast upon the shore of the realm were, except in certain privileged places, the king's property. Even to this day in England, Wales and Northern Ireland (there are certain exemptions in Scotland) 'Royal Fish, ie whales, porpoises, dolphins and sturgeons, whether dead or alive, belong the Crown except where the carcase is washed ashore or stranded within the limits of a Manor in respect of which the title to the Royal Fish has passed from the Crown to the Lord of the Manor. The liability for disposal or burial of carcases belonging to the Crown rests with the Board of Trade.'

Throughout this period the Basques of France and Spain were taking advantage of the abundance of whales in the Bay of Biscay and developing what is regarded as the first commercial enterprise in whaling history. The black right whale, known to the Basques as the *sarda*, was so numerous and appeared so close to the shore that the hunters were able to put out from the open beaches to pursue them.

It is generally accepted that Basque whaling reached its peak in the thirteenth century, but it did continue into the seventeenth century. Although Ambroise Pare's account of the Basque offshore whaling (1567) comes at a late date it does give a picture of their methods as they would be carried out over the whole of this period.

Because the artists were working from descriptions which were either inaccurate or misunderstood, on early woodcuts whales tend to have an absurd appearance, often combining features of both baleen and sperm whales. Fearsome stories were told of the 'sea monsters', and this gave rise to its being depicted as a frightening creature, as in this illustration of flensing dating from 1555.

Author's collection

Flensing the whale, 1590.

Author's collection

Celebrating Mass on the back of a whale, an engraving in 'Nova Typis Transacta Navigato Manacho', 1621.

Author's collection

Lookouts were established on cliffs or other vantage points and when a sighting was made some form of signal such as the beating of drums, the ringing of bells or the lighting of a fire was given. Whaleboats were launched as quickly as possible and were guided in the right direction by signals from the lookout.

The whale appears on the seals of several Basque towns and documents relating to the whale fishery, showing that it was an important part of Basque life. They gradually built up a commercial enterprise. Whales' tongues were sought after as a delicacy in the markets of Bayonne, Biarritz and Ciboure, while the blubber, after salting, was taken to various points inland.

Whaling grew in importance but as it did so the Basques found that the stocks of *sarda* close to the shore were diminishing either through 'overkill' or because the whale, with its highly developed sensory capacities, recognised the dangers and moved further to sea. Whatever the reason, the Basques were forced to sail further from the land to pursue the prey important to their livelihood.

They adapted the caravel, a proven ship, for whaling on the high seas. As long as a whale was caught within a practical towing distance of the shore the problems of dealing with the body did not arise. Once beyond this point the Basques had to devise other methods for disposing of the dead whale. If the voyage home was such that the blubber would arrive in reasonable condition they flensed (cut up) the whale alongside the ship and stored the blubber in casks. If, however, they were so far from home that the blubber would deteriorate before they reached port, there was another problem.

The importance of whales to the Basques is shown by the number of Basque towns which depict a whale on their seals and arms. They are Biarritz, 1351 (1); Motrico (2); Guetaria (3); Fuenterrabia, 1297 (4); Ondarroa (5); Le Queitio (6).

Author's collection

1

2

3

4

5

6

Whalers' implements used in the ice in the sixteenth century.

Author's collection

A whale attacked by killers.

Author's collection

Captain François Sopite solved this by erecting a try-works on the deck. This construction enabled a container to be placed over a fire so that the blubber could be converted into oil. Now it did not matter how far the caravels had to sail in pursuit of the whale.

These developments by the Basques were important and would be used by other nations when they entered the whaling trade in various parts of the world.

The Basques were the only Europeans hunting whales on extended sea voyages, and during the fourteenth, fifteenth and sixteenth centuries they had the monopoly of whale hunting on the high seas. They reached the Banks of Newfoundland in 1372 and were off the north-west coast of Iceland in the early part of the fifteenth century.

As explorers opened up new seas and reported numerous whales, other, more powerful countries turned their eyes to the possibilities of taking up the whaling trade. In spite of Sopite's development the

Basques found themselves unable to compete but, as the experts, their services were in demand with those maritime nations, notably England and Holland, who were still learning the art of whaling.

While these developments were taking place Basque whaling received its final, fatal blow when the Basque port of St-Jean-de-Luz was destroyed by the Spanish during the Thirty Years War. The occasional expedition sailed but the Basques never attempted to rebuild the industry.

While the Basques were masters of the whaling trade in Europe, the Japanese were developing ideas of their own. The inhabitants of Hokkaido, Japan's northern island, were killing whales by means of spears treated with poison made from aconite roots, in a similar manner to the natives of the Aleutian Islands, Kamchatka and the Kurile Islands. Whale hunting along the other coasts of Japan followed the more usual method of harpoon throwing from an open boat and at this time was carried out solely for the survival of the community.

A whale attacks a ship.

Author's collection

Hunting the whale.

Author's collection

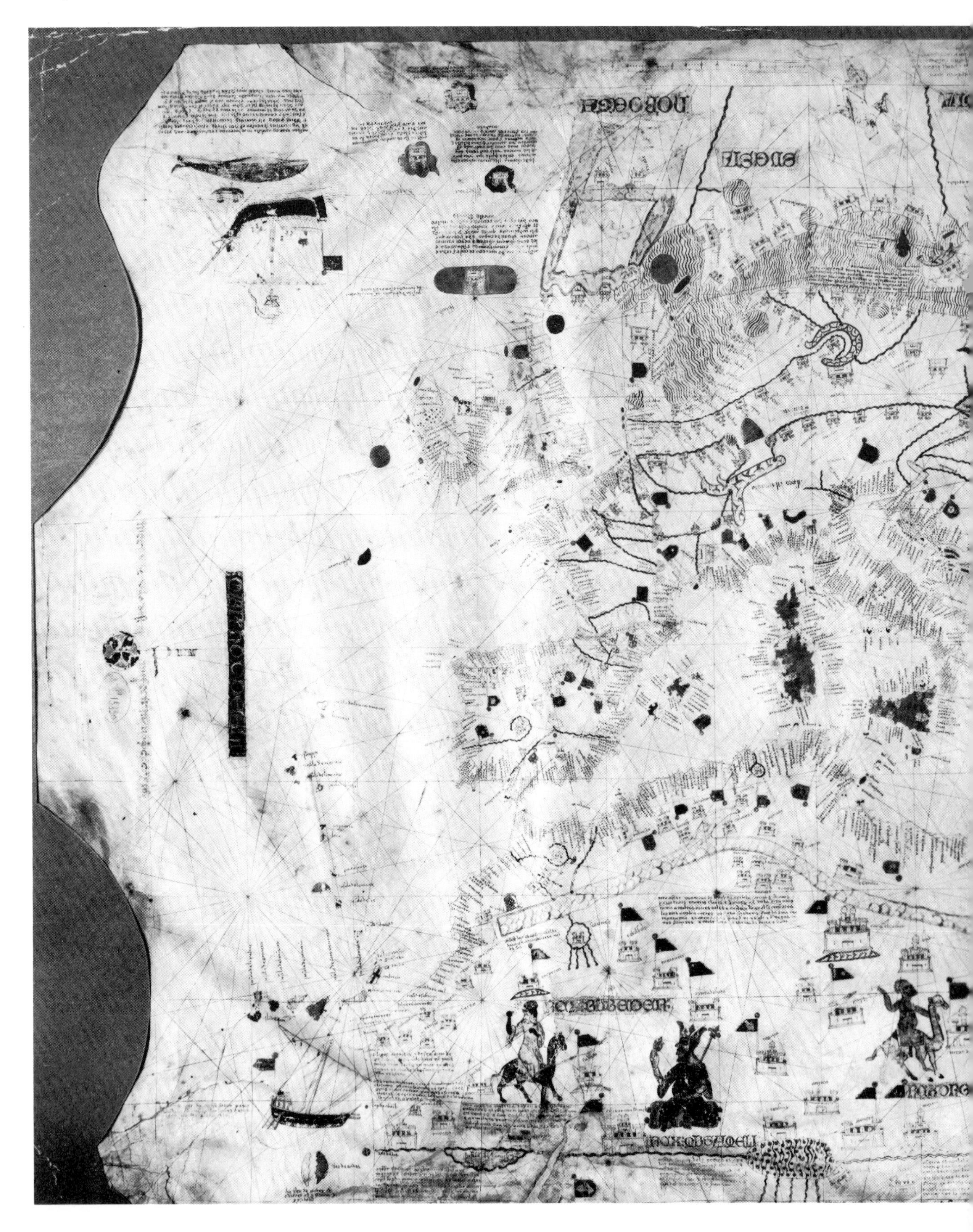

The 'Carta Catalan de Mecia de Veladestes' of 1413. As the Basques were the only people using caravels for whaling early in the fifteenth century, the fact that this map shows a whale and a caravel in the north-west corner, in the approximate position of Iceland, proves that the Basques were whaling far from their own shores.

Bibliothèque Nationale, Paris

At the beginning of the seventeenth century the Japanese developed an individual whaling technique. The hunters drove the whale into a net held by companion boats in shallow water. Entangled and exhausted by its struggles, the whale became easy prey.

Kendall Whaling Museum, Sharon, Massachusetts

In about 1600, the Japanese introduced a new technique for catching whales which is unique in the history of whaling. Nets, strongly constructed yet light enough to be handled by men in open boats, were used to entangle the whale. Six boats, working three nets, lowered them into water shallow enough to prevent the whale from escaping by diving underneath them. Twenty beater boats ranged behind and on either side of the whale and forced it towards the net-carriers. Once it was entangled the hunters took advantage of its helplessness, harpooned it and, when it was weak enough for them to get in close, killed it with lances. The whale was then towed ashore by more than ten boats in two ranks.

The method proved successful, but it was costly because of the high wastage of nets. This and subsequent developments put the whaling in a class which only wealthy merchants could afford. Consequently a well-organised and profitable whaling industry was established.

A factory was built as close as possible to the water's edge on a site with a suitable beach where the dead whale could be hauled in easily. There had to be room for numerous buildings for processing the whale and storing the products. To complete the ideal site there had to be elevated ground close by where lookouts could be stationed. The factory contained the try-works but there were also stores for nets, ropes, jars, oil tanks, whalemeat, tail flukes, bone and tendons, and workshops for the smiths and coopers. Each of the employees, who were housed in living quarters close to the factory, had a specialised job.

The main species taken in this type of whaling were the right, the humpback and the fin whales. Every possible use was made of the dead creature: the blubber was boiled for oil; the flukes and meat were salted and sold as food, some of it being exported; the bones were sawn and chopped, boiled, crushed and boiled again, to get as much oil from them as possible; and the residue was used as a fertiliser.

In spite of the capital outlay and everyday expenses, this method of whaling flourished in Japan for about 300 years.

Early Japanese whaling – the whale is drawn to the water's edge close to the factory where many people are engaged to cut up and dispose of the whale as quickly as possible.

Kendall Whaling Museum, Sharon, Massachusetts

Japanese whaling sampan.

Crown Copyright, Science Museum, London

Magdalena Fjord, Spitsbergen. This small but extremely beautiful fjord was used by the English in the days of Spitsbergen whaling.

Author

2 Spitsbergen

The English were aware of the Basques' successful whaling activities around Newfoundland and in 1594 the 35-ton *Grace* of Bristol was sent there to procure whale products, the first ship ever recorded to leave England for that purpose. There was no intention to hunt whales; the cargo was to be made up from stranded whales or wrecked whaleships.

It was not a successful voyage and, because whales had been reported much nearer home, the English had no need to send other vessels to the Newfoundland seas.

Early explorers searching for a North-east or North-west Passage reported sighting whales in the Arctic seas, especially off Spitsbergen. This information attracted the attention of the Muscovy Company which, in 1576, was granted a monopoly to kill whales, but it was slow to engage in whaling. Although Hull Merchants, ignoring the monopoly, sent their ships to hunt the whale off Iceland and the North Cape as early as 1598, it was not until interest was stirred again by Henry Hudson's reports of numerous whales off Spitsbergen that the Muscovy Company became more enthusiastic.

Even so the first whaling expedition did not sail until 1611. Realising that inexperience in whaling was a drawback to their English crews, the Muscovy Company recruited six Basques, experts in the trade, for this expedition. Instructions, detailed in *Purchas His Pilgrims*, were issued to Thomas Edge, company agent on the two-ship expedition, and these are probably the earliest for a whaling voyage ever recorded. Eight types of whale are described to help identification and Edge is told to treat the Basques 'very kindely and friendly' and to watch them very carefully so that he and his men can learn 'that businesse of striking the whale as well as they'. Apart from the experience gained, the expedition must rank as almost a total failure for both ships were lost although the crew and cargo were saved by a Hull ship.

In spite of this the Muscovy Company sent two ships the following year. They were more successful for they took seventeen whales yielding 180 tons of oil. This expedition, however, experienced the first

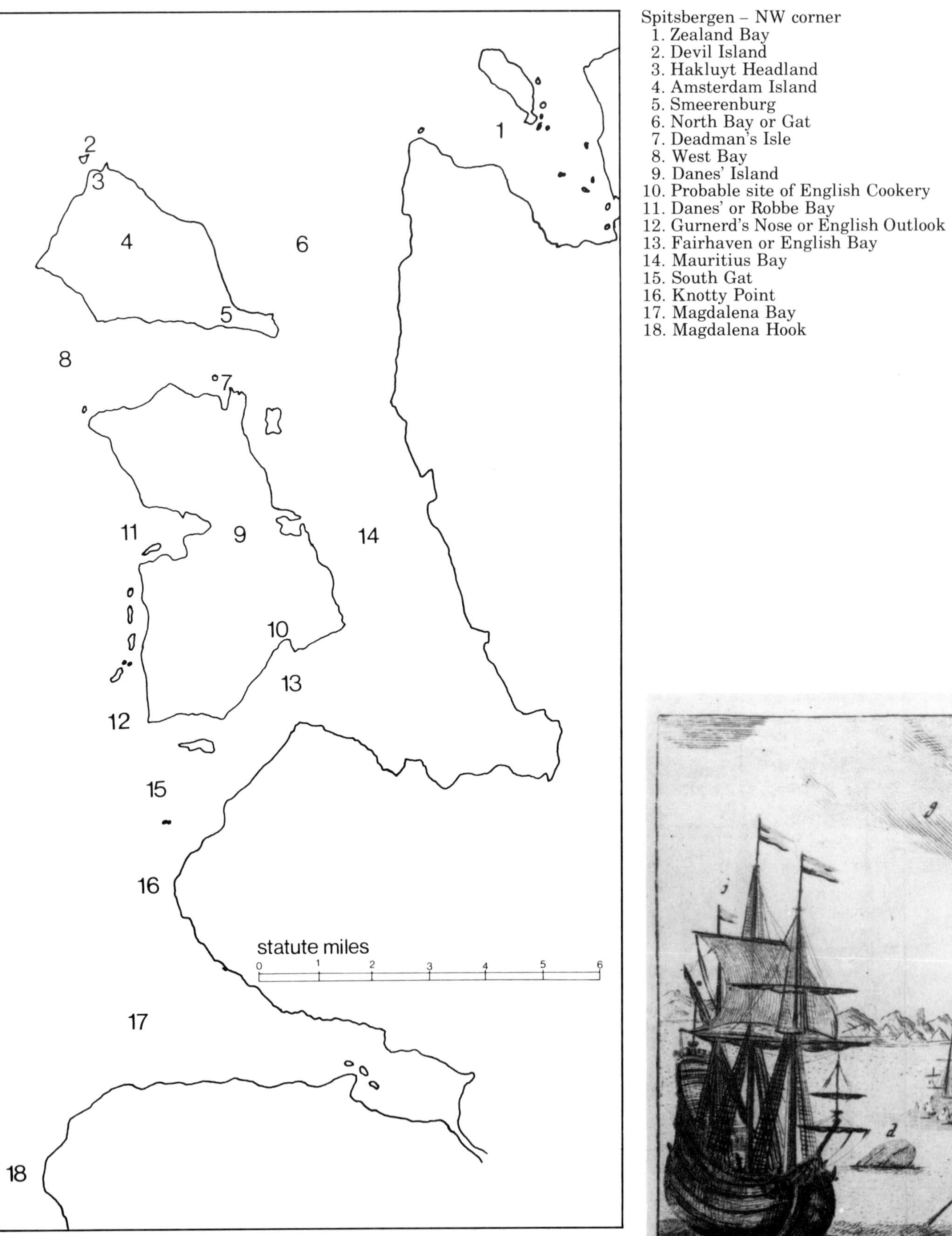

Spitsbergen – NW corner
1. Zealand Bay
2. Devil Island
3. Hakluyt Headland
4. Amsterdam Island
5. Smeerenburg
6. North Bay or Gat
7. Deadman's Isle
8. West Bay
9. Danes' Island
10. Probable site of English Cookery
11. Danes' or Robbe Bay
12. Gurnerd's Nose or English Outlook
13. Fairhaven or English Bay
14. Mauritius Bay
15. South Gat
16. Knotty Point
17. Magdalena Bay
18. Magdalena Hook

ripples of the disturbances which were to bedevil the whaling industry for many years. It met four ships on the whaling grounds which, according to the Company's monopoly, had no right there. Two of the ships were 'interlopers', the term given to any English ship sent whaling but not belonging to the Company. The other two ships were Spanish and Dutch.

To protect its monopoly, the Muscovy Company was granted a royal charter in 1613, but to enforce this monopoly was another matter. To give its ships protection against interference from foreigners and interlopers the Company sent the *Tigris*, a 21-gun ship, to accompany the 1613 expedition of six ships. A number of intruders were met and the *Tigris* exerted its authority by confiscating the whale products or exacting a portion of the catch. In spite of the presence of twenty-four experienced Basques among the crews the expedition took only thirty whales. This was too few for an expedition of this size which should have produced handsome profits but in fact incurred a loss of between £3000 and £4000. It would seem that the whaling was not pursued to its full extent and took second place to the constant distraction of clearing the area of interlopers.

The Dutch realised that it was no good going to the whaling grounds unless they went in force and were prepared to fight. Thus the stage was set in the Spitsbergen seas for a rivalry between the Dutch and English which persisted until the latter half of the seventeenth century. The fortunes of the English fluctuated and declined, whereas the Dutch, after a cautious start, went from strength to strength with only temporary set-backs and, as the size of their whaling fleet grew, their returns became greater.

Whalers at Spitsbergen: an engraving made in 1711 after Marten's voyage of 1671. The numerous activities of the whaling trade are clearly depicted.

National Maritime Museum, Greenwich

The Dutch knew that if their whaling industry was to prosper they would have to organise themselves to meet the English who, with greater experience on the whaling grounds, might be expected to expand in order to keep ahead of any rivals. A number of Dutch merchants put aside their jealousies and, in 1614, formed the Noordsche Company. The object of creating such a monopoly was not only to compete with the English Muscovy Company but also to make use of the ensuing financial advantages for the development of the Dutch whaling industry on a large scale.

When the Dutch whaling fleet of fourteen ships sailed in 1614 it was escorted by four men-of-war, each of 13 guns. Anticipating such a move, the Muscovy Company sent thirteen ships and two pinnaces. When the two fleets met, common sense prevailed and they came to an agreement whereby the English whaled at Bell Sound, Ice Sound, Fair Foreland and Fairhaven, while the Dutch could hunt from any other place unless a whaler belonging to the Muscovy Company was already there. This compromise could have been the basis for a permanent truce at Spitsbergen resulting in lucrative whaling for both sides, but it was not to be, and the situation was made more complex in 1615 when the Danes advanced a claim on Spitsbergen and its fisheries, supporting it by sending three warships to the area. However, their attempt failed, as did others in the following years.

'Whale catchers' by A Salm, 1700 – Dutch whaling at Spitsbergen.

By courtesy of the Maritiem Museum 'Prins Hendrik', Rotterdam

'Whaling' by van der Laan. The Dutch were the leading whalers in Europe throughout the seventeenth century. At the height of their industry they had as many as 200 ships at Spitsbergen.

By courtesy of the Maritiem Museum 'Prins Hendrik', Rotterdam

The rumblings of the dispute between the English and the Dutch continued and more difficulties for the Muscovy Company arose when King James granted to Sir James Cunningham and his heirs and associates the right to trade in the East Indies, the Levant, Greenland, Muscovy and all other countries and islands in the north, north-west and north-east. Thus the new Scottish East India Company was granted the right to hunt whales if it desired.

A decline in the Muscovy Company's finances – even though 150 whales yielding 1900 tons of oil were killed in 1617 – brought the two rivals together, and the Scottish East India Company agreed to make a loan to the Muscovy Company provided that the whale fishery was carried on jointly for eight years.

Throughout this time negotiations were going on between the English and the Dutch over the rights to Spitsbergen and its whaling grounds. The Dutch claims were based on the right of discovery of Spitsbergen by Barents, Heemskerke and Ryp in 1596. The English supported their claim by the fact that they were the first to whale in the area.

The Dutch seemingly had the backing of force when the first joint effort of the Muscovy Company and the Scottish East India Company came up against a strong force of twenty-three ships. A number of Englishmen were killed and one English ship was taken to Holland. The English were unable to protect their interests by sending an efficient naval escort, since at this time the Navy was not in a very healthy state – only twelve ships were serviceable and these were required elsewhere. The joint venture ended in disaster with a financial loss of £66,000 and both companies made strong complaints about Dutch intervention.

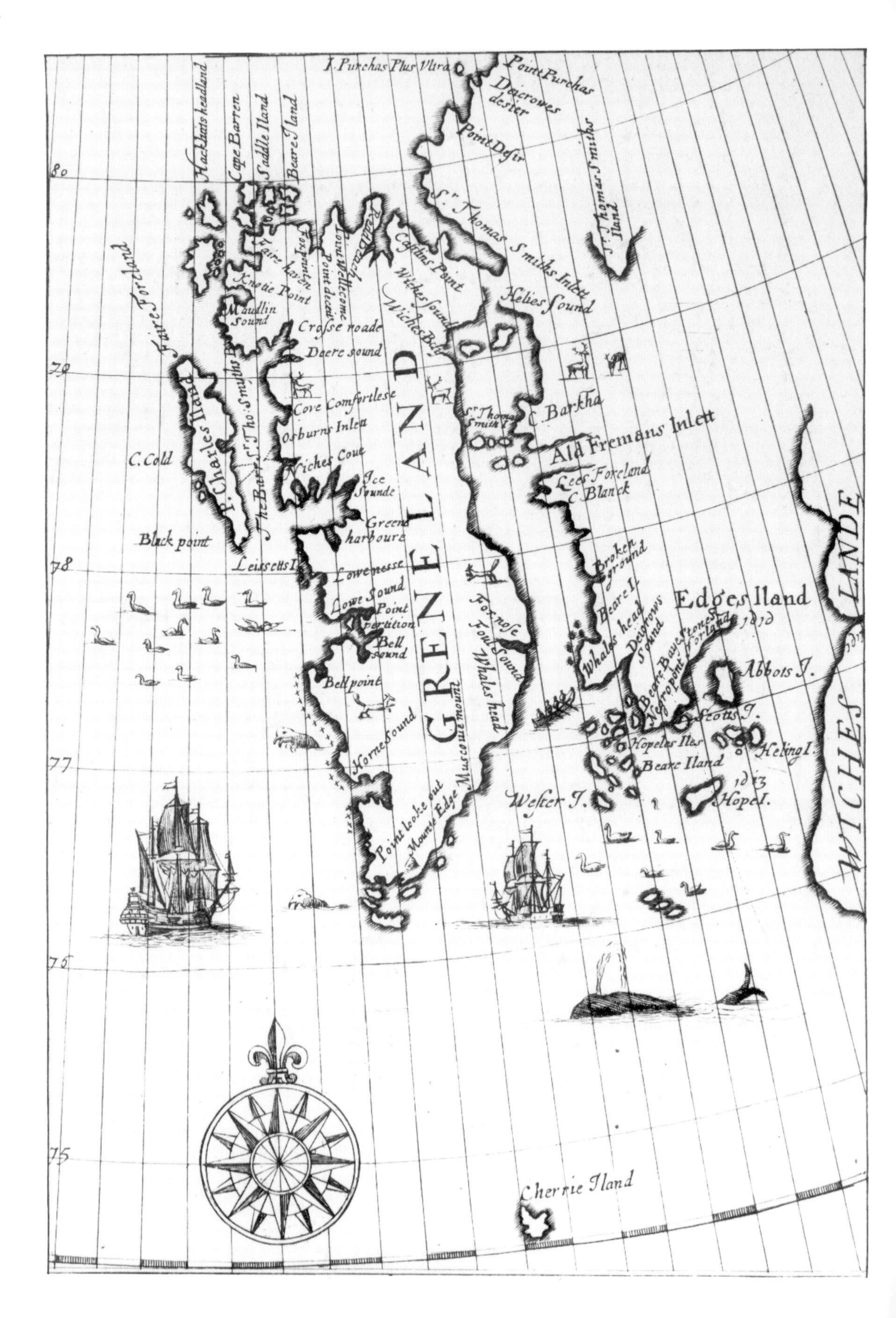
I. Purchas Plus Vltra
Point Purchas
Desirowes desier
Point Desir
Hackluits headland
Cape Barren
Saddle Iland
Beare Iland
S.r Thomas Smiths Iland
S.r Thomas Smiths Inlett
Faire Foreland
Faire haven
Knotie Point
Maudlin Sound
Point deceit
Coffins Point
Wiches sound
Wiches Bay
Helies sound
Crosse roade
Deere sound
GRENELAND
Cove Comfortlese
Osburns Inlett
Wiches Coue
S.r Tho: Smyths Bay
P. Charles Iland
The Barr
C. Cold
Ice Sounde
Greene harboure
Black point
Leissetts I.
C. Barkha
S.r Thomas Smiths I.
Ald Fremans Inlett
Lees Foreland
C. Blanck
Lowenesse
Lowe Sound
Point Pertition
Bell sound
Bell point
Horne Sound
Foxe nose
Foul Sound
Whales head
Muscovie mount
Broken ground
Beare I.
Whales head
Edges Iland
Deuiltows Sound
Negropoint Forland
Beare Bay
1616
Abbots I.
Scotts I.
Hopeles Iles
Beare Iland
Heling I.
Wester I.
1613
Hope I.
Point looke out
Mount Edge
WICHES LANDE
Cherrie Iland
80
79
78
77
75

Although the Dutch gained the upper hand during the 1618 season their government did not want to prolong the quarrel and sent representatives to England. King James would not give up his claim to Spitsbergen but agreed not to enforce it for three years, during which time both countries were to carry out their whaling without interference from each other. In return the Dutch were to hand back within three months goods which they had confiscated or else pay £22,000 within three years.

However, the Dutch put forward three proposals: the bays and fishing stations should be shared by all nations; the whaling should be pursued by Dutch and English vessels of equal size and number; and the island should be equally divided between the two nations. The English were stubborn, but finally agreement was reached and a division made of the bays and harbours, which were to be considered the independent possessions of those to whom they were allocated.

The English had the first choice, followed by the Dutch, the Hamburgers, and the Biscayans. The principal ice-free southern bays, Bell Sound, Preservation or Safe Harbour, Horizon Bay, English Harbour, and Magdelena Bay were chosen by the English. Among the Dutch sites was Amsterdam Island which was to become an important whaling base. The Danes took up positions between the English and the Dutch, while other claimants moved further north.

The Bay Fishery, as the Spitsbergen whale-fishery was known throughout its first thirty years, thus offered whaling facilities to various nations, but only the Dutch profited in a big way.

The English efforts proved unprofitable and in 1620 the Muscovy Company sold its whaling assets and liabilities for £12,000, but the new owners fared little better and English whaling became only a spasmodic affair for a number of years.

Throughout this time all the hunting and processing equipment had to be transported to Spitsbergen each season, but to save time offloading and setting up, it became the practice to leave it there as a semi-permanent shore station. Even more time could be saved if men wintered in Spitsbergen and prepared everything for the arrival of the whalers in spring.

With this in mind the Muscovy Company obtained permission to take condemned prisoners to Spitsbergen with the promise of a pardon if they would fulfil this role but, willing though they were, the prisoners were so terrified on seeing the bleak and desolate landscape of barren jagged mountains through which huge glaciers carved their way to lonely fjords that they chose to relinquish their pardon and return to England. No amount of persuasion could get them to face a winter of darkness, intense cold, storms and fog in such an inhospitable-looking country. When the expedition returned to England, the Company interceded for the prisoners, who received a pardon despite the fact that they had not carried out their obligations.

The summer months, however, offered the opportunity to the whalers to gain handsome profits. While the English, plagued by disputes among themselves, by bad management and by the country's poor economic condition for many years prior to the Civil War, failed to seize the chances, the Dutch, with the bigger European market for their oil, 'thought big' and acted quickly.

Map of Spitsbergen from *Churchill's Voyages*. This early seventeenth century map of Spitsbergen is wrongly marked 'Greenland' because for many years Spitsbergen was thought to be part of Greenland.

By courtesy of the Minster Library, York

The building of boiling-houses, warehouses and cooperages at their allotted bays involved considerable initial expense, but led to greater financial gain in the long run. The number of men and quantity of stores were adjusted according to the size of the ships, and any extravagances which would lower profits were eliminated. They also adopted a system of 'bottoming', by which tradesmen supplying a whaling expedition with goods were not paid in cash but were encouraged to place them against the success of the voyage, taking a profit according to the goods or service provided. This method of setting up an expedition meant that the owners of whaling vessels were not faced with a huge loss in a poor season.

In 1623 the Dutch decided that they would establish a big settlement at Smeerenberg on Amsterdam Island. They shipped massive quantities of supplies, including all the necessary timber (there were no trees on Spitsbergen) to build and maintain a small town during the whaling season. Each chamber of commerce which made up the Noordsche Company had its own site and 'tents', as they were called, where their equipment was housed and the dead whale processed when it was hauled ashore.

'Hunting the Whale at Jan Mayen'.

Author's collection

Dutch and German whalers in the Arctic.

Author's collection

Whales were plentiful close to the shore and once a sighting had been made shallops manned by six men put off from the shore in pursuit. When they were close enough the harpooner threw his weapon, which was not the means of killing the whale but merely the way the boat was attached to the prey. From then on it was a case of playing the whale, running out more rope if it sounded so that the boat was not dragged under, running with the whale if it swam on the surface, judging the right time to take in the rope so that the boat would be taken nearer the whale to enable the harpooner to use his lance to kill the whale, and all the time being alert to the possibilities of the whale coming up under the boat or the flukes dealing a pounding blow of destruction.

Once the whale was dead it was towed back to the shore. It was left for a day or two then the fat was taken off and brought ashore to have the oil extracted. A watersideman removed the fleshy parts and cut the blubber into pieces weighing approximately 2cwt each. These were barrowed by two men to a platform on a stage where a stagecutter reduced the pieces to slices about 1ft long and 1½in thick. After being put into a 'slicing cooler' each slice was cut into smaller lumps about ¼in thick and 1–2in in length, by up to half a dozen men each working at his own chopping block. The blubber was put into a 'chopping cooler' which held 2–3 tons of the substance, and from there it was taken by a 'tub-filler' to a copper in which it was boiled. Two coppermen were in charge of the operation and it was their job to transfer the oil, using long-handled copper ladles, into a 'fritter barrow' which had a wooden grating at the bottom through which the oil was strained. From here it passed through three cooling tanks of about 5 tons capacity. These were placed at different levels in front of each other so that when the oil reached a certain height in the first it automatically ran into the second and so on into the third from which the casks were filled. The casks were taken to the ship by forming them into rafts of twenty. Meanwhile the cleaning of the whalebone, the long fibrous strands which hung in the mouth, had been going on and, after packing it into bundles of sixty pieces, it was transported to the ship in a longboat.

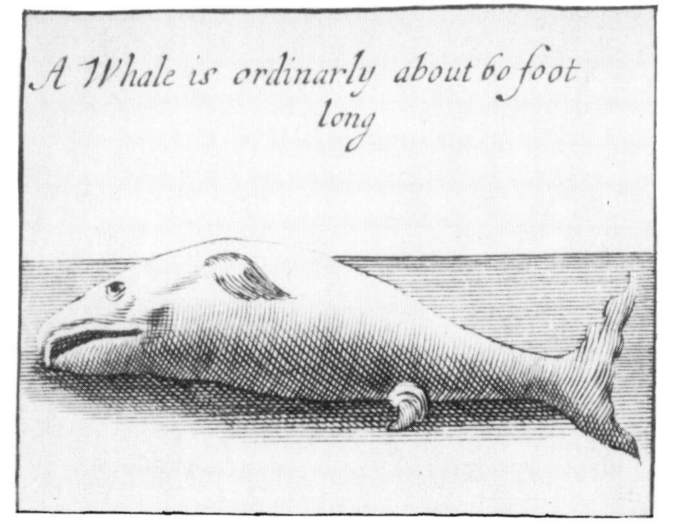

An early seventeenth century description of the stages of hunting in the Spitsbergen whale-fishery.

By courtesy of the Minster Library, York

The advantage of the shore station was that while some of the men hunted the whale others did the flensing and operated the cookeries. Uninterrupted whaling meant bigger catches and therefore more profit. This system meant a doubling up of crews, so 200–300 ships – a number not uncommon in the heyday of Smeerenburg – each with an average crew of 30, gave a total of between 12,000 and 18,000 men at the settlement during the whaling season.

Apart from the blubber cookeries, workshops and warehouses, there were also bakeries, a store, a church and inns. Smeerenburg boomed during the season but when the whalers left and the Arctic night took over it became a ghost town lost in the snow, with only the shrieking sound of the freezing wind and the boom of the treacherous winter sea to contest the silence.

In spite of the huge outlay and the fact that it was within 10° of the North Pole, Smeerenburg proved not only to the Dutch but also to other interested nations the feasibility of shore stations, though these were not to come into prominence for another 200 years.

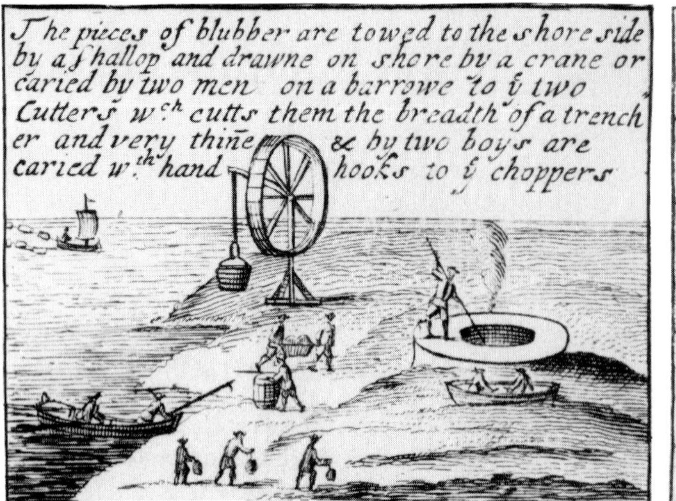

During the winter of 1633–34, the Dutch attempted to colonise Jan Mayen and Spitsbergen, not only to consolidate their claim but in order to protect property left there for the following season, since in 1632 supplies on Jan Mayen had been looted by Basques in retaliation for being driven away from Spitsbergen. No doubt the Dutch were encouraged in their project by the fact that eight Englishmen had successfully wintered on Spitsbergen in 1630–31. They had become separated from their whaler, the *Salutation*, and, thinking they had perished, the captain sailed for England. When the whalers arrived the following year they were suprised to find the men they thought dead very much alive, the first men to spend a winter on Spitsbergen.

The Dutch venture, headed by van der Brugge, was well planned but the seven men sent to Jan Mayen died, and although the seven on Spitsbergen survived, the conclusion reached was that conditions were too extreme to warrant a permanent settlement and the whalers continued to come just for the whaling season.

The French were regular visitors to the Spitsbergen seas on a small scale but were continually challenged by the Dutch. Annoyed, the French eventually challenged the Dutch rights in Spitsbergen and forbade the Basques to serve on Dutch ships. This did not deter the Dutch, who had by then acquired experienced whalers from another source. Men from the Frisian islands of Sylt and Fohr, where there was little opportunity to earn a livelihood, often served on foreign vessels, many of them on whaling voyages, where they learned the trade and became the sort of experts welcomed by the Dutch.

In 1626 a change in the manner of whaling was forecast when two whalers from Zaandam became the first Dutchmen to kill a whale in the open sea. Whales were still plentiful close to Spitsbergen and continued to be the basis of the industry for a good many years but there were indications that their numbers were declining. Large-scale whaling by the Dutch plus the catches, though smaller, by other countries affected the whale population. Man had to turn elsewhere and, just as the Basques had done when the whale population in Biscay had declined, the nations whaling at Spitsbergen had to turn more and more to the open sea.

In 1639 two ships from Amsterdam hunted the whale in the open sea between Spitsbergen and the North Cape and three years later ships from Zaandam were regularly engaged in open-sea whaling. At first the blubber was boiled at sea in the manner used by the Basque, Sopite, or on some nearby shore, but after 1642 it was more usual to take it back to Holland, where it could be processed under more favourable conditions, resulting in better quality oil and less wastage of byproducts.

The Noordsche Company lost its monopoly: in 1645 Dutch whaling was thrown open to all who desired to pursue the trade, and the number of ships going to the Spitsbergen seas increased. However, by this time the number of whales being caught near the shore made it impracticable to keep Smeerenburg as a shore station. By 1645 its deserted buildings and the graveyard on nearby Deadman's Island were all that remained of a once thriving town. Whaling in the northern seas was about to enter a new era.

3 American Beginnings: Dutch Supremacy

The history of whaling is closely connected with exploration and the opening up of North America is no exception. The first explorers and arrivals reported seeing many whales but the first settlers, having little if any knowledge of pursuing and killing the whale, were content to make use of those stranded on the beach. They saw the coastal Indians hunting the whale from their slender canoes, in a manner basically similar to early peoples elsewhere across the world.

It was fairly common practice for the Indians to use 'drugs', pieces of wood about 2ft square with a stick through the centre which was attached by a rope to a harpoon roughly fashioned out of bone or wood. These drugs served the purpose of slowing down the whale; several harpoons were embedded in the whale for greater effect, thus enabling the Indians to get close enough to kill the whale with lances.

Observing the Indians whaling, it was natural for the settlers to copy their methods and enlist their aid instead of relying on drift whales. No doubt realising that by adapting their heavier row-boats and implements even better results could be achieved, they began to use a stronger harpoon, with a metal head, and a more serviceable lance.

Opinions differ as to whether the settlers followed the Indian use of drugs or the European method of 'fastening on' to the whale by harpoon-line. No doubt both methods were used, and as the practice of fastening on became more widely known it largely superseded the earlier use of drogues for, though more dangerous, it enabled the hunter to keep in closer touch with the whale and make an earlier kill. Nevertheless the American whalemen did not abandon the drogue completely; it became part of the equipment of every whaleboat, for use when the whale was making its escape with the line still attached to it. From the time when the white settlers first hunted the whale from North American shores, Indians, highly thought of for their skills, formed part of the crew and were to do so throughout most days of American whaling.

Offshore whaling – flensing the whale on the beach, Long Island, USA.

Author's collection

Most probably the first whites to pursue the whale from the American coast were the people of Southampton, Long Island, some time between 1645 and 1651. The crew would sail up and down the coast searching for whales for periods up to three weeks, coming ashore to camp at night. Later, lookout points were erected along the coast with huts nearby for the crews, so that the boats could be launched with the minimum of delay once a sighting had been made. The lookout post was a flat wooden perch set on top of a pole, and a man was specially appointed to scan the sea. In 1651 John Mulford was engaged as a look-out by the whalers of Easthampton, Long Island. His son Samuel, who was six at the time, was to become a champion of the whalemen's cause against injustice.

Whales were being hunted by the people of Martha's Vineyard in about 1652 and from the coast of Cape Cod before 1670. By 1688 whaling had grown to such an extent that Secretary Randolph in a report to England said 'New Plimouth Colony have great profit by whale killing. I believe it will be one of our best returns now bever and peltry fayle us.'

Throughout this period Nantucket seems to have ignored the plentiful supply off its coast, but eventually the islanders decided that the sea must be the basis of their livelihood and that the whale could become the mainstay of their economy. They knew little or nothing about hunting at sea or about the organisation of the whaling industry so, in 1690, they brought Ichabod Paddock from the whaling community of Cape Cod to instruct them.

Paddock divided the south shore of the island into four sections, each between 3 and 4 miles long. In each of these sections a lookout point was erected and manned by one of the six men appointed to the section while the other five lived in a hut nearby. When a whale was sighted all the crews joined in the pursuit and helped each other with the kill, cutting up and trying-out.

Nantucket's communal efforts led to a successful enterprise unique in the annals of whaling. The small island was ideally situated and its people were a closely-knit community of hardworking Quakers who had come there to escape the Puritan attitudes to their way of life which existed on the mainland – and once the whaling trade was established they remained at it all their lives. The whalemen had a certain standing in the community whose livelihood depended on them. As there were plenty of young men eager to serve on whaling expeditions, it was not uncommon for a whaler to give up going to sea when he was forty and take a shore job in the industry in a craft he had learned at an early age before going to sea.

At this time the whaling business in Europe was conducted on a 'master–servant' principle whereby the employee was paid for his services while the employer took the remaining profit. The Nantucketers, however, working on a communal basis, distributed the proceeds in proportion to the amount, either in labour or in kind, each person put into the enterprise. By the end of the seventeenth century, a growing whaling industry was established along the New England coast where the try-works glowed, manufacturing oil for home consumption.

The Dutch in the Greenland whale-fishery. This painting of 1843, by J Mooy, shows the Dutch *Frankendaal*, Captain Maarten Mooy, father of the painter (second from right); the *Groenlandia* (centre); and the *De Jager* (left).

Author's collection

Flensing the whale. The caption points out the windlasses used to haul the whale on shore.

Author's collection

Across the Atlantic, in the northern waters of Europe, the Dutch still reigned supreme, but seventeenth century wars hindered their whaling trade in a number of ways: there were seasons when the whaling fleet was forbidden to sail; such experienced men as the whalers were in great demand for the Dutch navy; and whaling men started working for foreigners. In 1661 the Dutch government forbade this practice as it was harming the economy and helping rival whaling countries. Apart from these occasional set-backs the Dutch whaling industry was booming and it was given an additional boost in 1675 when the government allowed Dutch whalers to bring their products into Holland tax-free while the tax on those imported by foreigners was increased.

Deep-sea and offshore whaling were carried on side by side in the Spitsbergen area for a considerable number of years until hunting the whale from the shore became uneconomical. Once out of convenient reach of Spitsbergen, a parent ship had to serve as a base for the shallops. At first a dead whale was flensed on the nearest shore or more often on an icefloe, but before long it became customary for the ship to anchor to an icefloe and drift south with it during the season while open boats hunted from the parent ship. The boats brought the dead whale back to the floe where it was cut up and the blubber stowed in casks before being taken home to be tried out. An alternative method was to fasten the whale to the side of the ship and flense it there, a practice which was to become universal as whaling extended across the world.

Towing the whale – Dutch whalers off Spitsbergen.

Author's collection

Lancing the whale – Dutch whalers off Spitsbergen.

Author's collection

A Dutch whaler at work in the Arctic. From an engraving by D Moy used in J A van Oelen's book on whaling, published in Leyden 1683–84.

Author's collection

While the Dutch whaling industry prospered, that of England was still bedevilled by disputes over monopolies, until eventually in 1672 the whaling trade was made free for all. Whale products were exempt from tax if brought in by English ships (a small tax was imposed on those brought by colonial vessels and foreigners paid a heavier duty), but there was no immediate upsurge in the English whale trade. In 1693 the London Greenland Company was formed but by 1704 its total capital had been lost. The misfortunes of this company deterred others from trying and the trade languished. During the same period the Dutch were particularly successful and the German whaling industry was in better shape than that of the English.

The English, having preceded everyone else into the trade, had every opportunity to build up an equally flourishing industry. English seamen were among the best in the world and they quickly learned the art of whaling from the foreigners who had to be employed at the outset. It seems that the main trouble was in the failure to grasp the opportunity. Once the Dutch saw the possibilities they thought ahead and organised accordingly; they 'thought big' while the English did not. The more ships hunting, the better the chances of making a profit, for the unsuccessful ship would be covered by the others. The English did not send sufficient ships really to make the hunting pay – when this situation was rectified in the late eighteenth century the success of the industry improved dramatically.

So, as the seventeenth century drew to a close the Dutch were predominant in European whaling; across the Atlantic the North American colonists were chiefly engaged in offshore whaling, with whales so plentiful there was little to tempt them into deeper waters. But early in the eighteenth century an event, which took place almost by accident, altered the course of whaling, especially in the North American context.

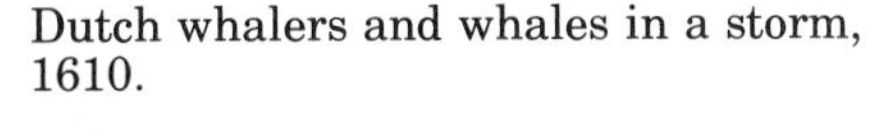

Dutch whalers and whales in a storm, 1610.

Author's collection

4 Wary Expansion 1712-83

A whaling scene of 1763. The initials J R on the barrels refer to Joseph Russell who founded New Bedford and probably started whaling there in 1755. The scene shows blubber being prepared for boiling.

From 'The Marine Mammals of the Northwestern Coast of North America' by Charles M Scammon

In 1712 a Nantucket captain, Christopher Hussey, blown off course in a small whaling sloop, found himself close to a school of sperm whales and succeeded in killing one. This incident, simply recorded, proved of great importance.

The sperm whale was known to the people of New England from the occasional one found stranded on the beach. Although they were aware of the superior quality of its oil and its valuable reservoir of spermaceti, the whalers had never pursued it because there were plenty of right whales close to the shore which fulfilled their needs. This plentiful and convenient supply outweighed the risks and uncertainty of undertaking deep-sea whaling. Offshore whaling had spread to all the main communities along the New England coast and reached its peak in the 1720s. The move from offshore to deep-sea whaling was gradual; both were carried on side by side for many years.

By 1726, 25 sloops of between 30 and 50 tons were hunting the sperm whale from Nantucket while there were 28 boats still working the offshore fishery. Nantucket whaling was building up but the whaling communities soon came into conflict with the authorities when taxation was introduced into the North American colonies.

In New York, Governor Robert Hunter ordered the whalers to pay a tax of one-twentieth of all oil and bone taken, whether from drift whales or those captured by boats. The duty had to be paid in New York which was approximately 100 miles from the whaling grounds. When the whalers ignored the order, a writ was issued in 1711 directing officials to seize all whales and compel the whalers to pay the share of the products due under the ruling.

Samuel Mulford was chosen to present the whaler's case before the General Assembly. He maintained that American whalers had the same rights under law as British whalers, and that the Governor, by imposing a tax, was acting contrary to the law which encouraged whaling. In order to thwart the whalemen's aims, the Governor made sure of gaining a majority in the General Assembly by creating new districts in which he was certain of support. Mulford was prosecuted and convicted

Cutting up the whale.

Author's collection

for refusing to pay the tax. Determined to take the matter up personally with the government in England, he left America secretly and was successful in his plea. Governor Hunter was ordered to withdraw the tax on whale catches.

As the whaling industry expanded, the New Englanders were producing oil in excess of their needs and found a ready market for it in England, where the whale-fishery at this time was virtually at a standstill. It was a peculiarity of this export trade that much of it was handled by the larger, commercial ports which had little or no interest in the actual whaling itself. Nantucket oil was, at one time, sent to Boston and then on to England, but the Nantucketers realised that this was losing them money and from then on sent their oil direct to England, where they found they could buy goods more cheaply than on the American mainland.

Whaling was encouraged in many ports along the New England coast and ships began to range further and further from home. Some ports specialised in some particular aspect of the trade – Provincetown sent her ships to Davis Strait and neighbouring seas to hunt the right whale – but generally the American whalemen preferred the less risky, more temperate seas inhabited by the sperm whale which provided a more valuable cargo. The pursuit of the sperm whale brought a boom to the American whaling industry and its allied trades, drawing the vessels across the world and giving crews sightings of the great rorquals of the south.

Some of the principal whaling ports in the USA.
1. Newburyport
2. Gloucester
3. Salem
4. Lynn
5. Boston
6. Dorchester
7. Quincy
8. Plymouth
9. Provincetown
10. Wellfleet
11. Falmouth
12. New Bedford
13. Providence
14. Stonnington
15. Mystic
16. Groton
17. New London
18. Bridgeport
19. New York
20. Greenport
21. Jamesport
22. Mattituck
23. Sag Harbor
24. Easthampton
25. Southampton

Geographical features
A. Long Island
B. Cape Cod Bay
C. Buzzards Bay
D. Martha's Vineyard
E. Nantucket Island
F. Cape Cod

At this time sloops usually hunted in pairs, one in support of the other when an attack was made. Each sloop carried a whaleboat, and once the whale was harpooned it was allowed to run, hampered by drogues, to be picked up later. The sperm whale, however, was swifter than the right whale and, so as not to lose it after a strike, it became standard practice to fasten on, which gave rise to the term 'Nantucket sleigh ride' being used to describe a boat pulled through the sea by a whale.

Meanwhile European whalers were still sailing the northern seas as far as the edge of the ice, and in London the numbers in 1721 were listed as 251 Dutch, 90 Danish, 55 from Hamburg, 24 from Bremen, 20 from ports around the Bay of Biscay and 5 Norwegian.

In England, the South Sea Company, which had been formed in 1711 with the aim of restoring credit and demolishing the floating debt, showed an interest in whaling. The arguments put forward by Henry Elking for pursuing the trade impressed Sir John Eyles, the sub-governor, and a proposition was put to the company in 1721. As reasons for the failure of previous English excursions into the whaling trade, Elking cited lack of knowledge on the proper way to conduct a whaling enterprise. Foreigners, skilled in the trade, had to be taken on and, in short supply, demanded exorbitant wages. The captains were on a fixed wage, whereas it would have given them greater incentive to press home their hunt for whales if they had been paid according to the value of the cargo they brought home. The whales which were brought back

The hazards of the whale-fishery.

Author's collection

were treated in badly managed cookeries, which affected the quality of the products. Instruments and implements were not cared for and many had to be replaced every season. Elking recommended that as well as putting these points right, the company should engage captains of experience able to judge all manner of conditions to the advantage of the company. He also stated that the ideal number of men for a ship of 300 tons and carrying six shallops was 42 or 43.

The directors were convinced and, following the encouragement given by an Act of Parliament by which all produce from the Greenland Sea brought in by British ships whose commanders and least one-third of whose crews were British subjects would be exempt from duty during the seven years from Christmas 1724, ordered twelve ships, each of 306 tons, to be built and ready to sail in spring 1725.

There were no experienced Englishmen to fill the specialised jobs of the whaling trade; a few Scots, who had gained the necessary knowledge while serving with the Dutch, were tempted to join the expedition, but most of the skilled men were from the North Frisian Island of Fohr. Although the catch should have shown a profit, high costs swallowed the money. These were due partly to extravagances in equipping the expedition, but chiefly to the high wages paid to the skilled foreigners who also had their passage paid to and from their homes. Until the British acquired specialised whaling experience, their expeditions were at a disadvantage.

'Whalers in the Arctic' by R Willoughby

By courtesy of Kingston-upon-Hull Museums

The Dutch first sailed into Davis Strait in 1719, and the British took steps to encourage their whalers to go there. The exemption from tax of catches made in the Greenland Sea was extended to include those made in Davis Strait from 1726 onwards. It was calculated at this period that for a voyage to be successful each ship must catch three whales. It was also estimated that one good year could make up for six bad ones, but unfortunately eight years were bad and in that time the South Sea Company's ships, over 20 per season after 1725, failed to average one whale per ship.

So the venture ended in disaster, but it would be unfair to write it off as a total failure. It revived interest in the whaling trade and laid the foundation for more successful ventures to come. The government was stirred into taking action to set the industry on a firmer basis. In 1733 the first Bounty Act was passed, allowing a bounty of 20*s* per ton on all ships of 200 tons or more fitted out in Britain for the purposes of whaling. However, it had little effect, and neither did the fact that the bounty was increased to 30*s* in 1740; in the period prior to 1749 no more than six whaling ships sailed in a season.

The government did not abandon its efforts to stimulate the whaling trade and in 1749 a Bounty Act was passed which was partially instrumental in reviving the industry, as more merchants were attracted to look into its prospects by an increase in the bounty to 40*s*. This, coupled with the growing demand for oil and bone throughout the country, brought more ports into the whaling trade which began to expand quickly.

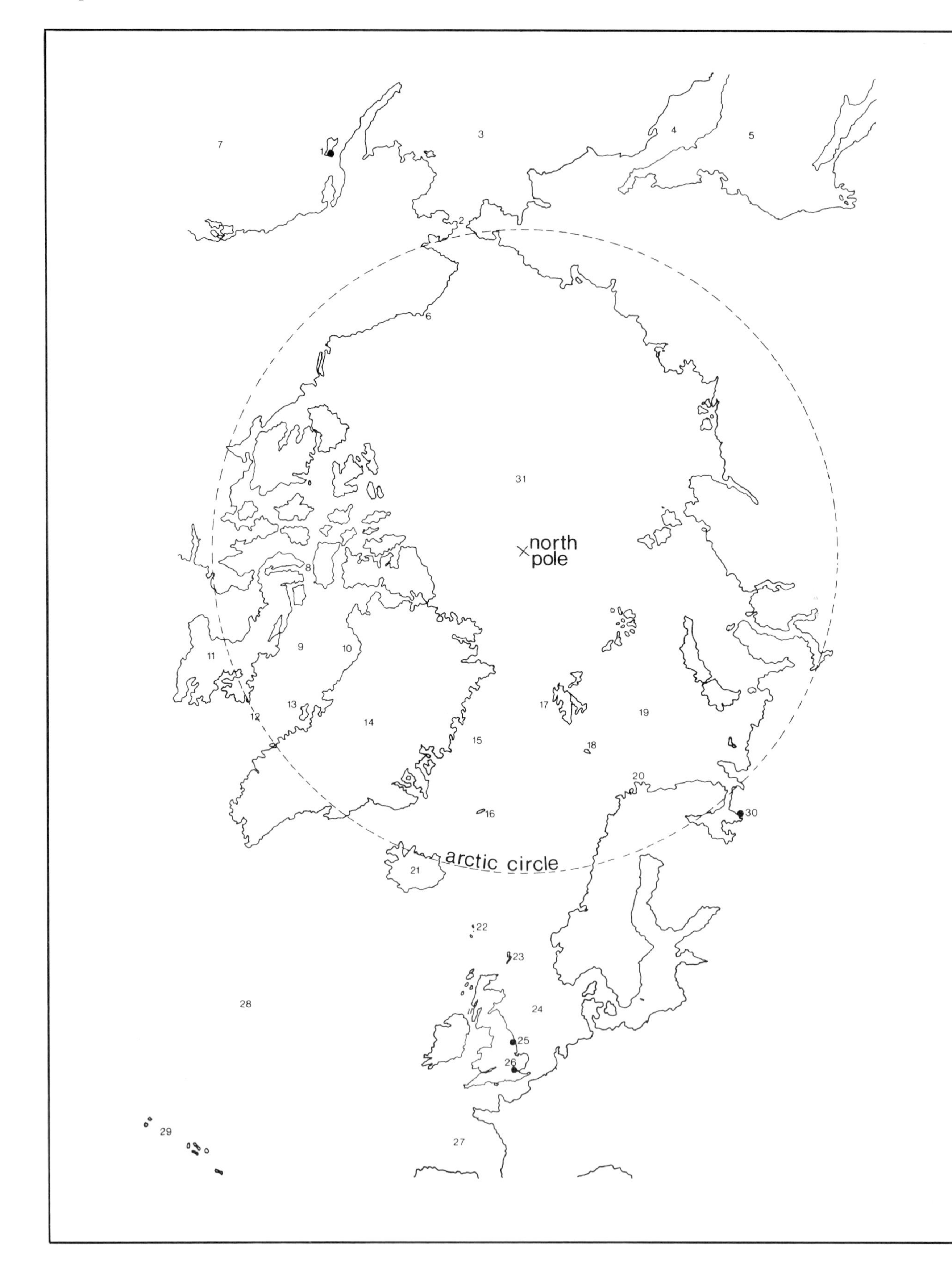

1
2
3
4
5
6
7
8
9
10
11
12
13
14
15
16
17
18
19
20
21
22
23
24
25
26
27
28
29
30
31
north pole
arctic circle

The dangers of the whale-fishery. This picture was originally produced in W Scoresby Junior's *Account of the Arctic Regions*, and shows the not infrequent fate of a whaleboat during the hunt.

By courtesy of the Whitby Literary and Philosophical Society

One of the most significant effects of the 1749 act was that it stirred the interest of the Scots and, although their whaling industry was to fluctuate, the Scottish ports eventually became the backbone of the British whaling trade and helped to prolong its life.

By 1753, 35 English ships, 25 of which were from London, sailed for the Arctic, while 14 ships left Scottish ports to hunt the whale. This was the year when Whitby, later to become an important whaling port, sent its first ship to the Greenland seas, and the following year saw Hull re-enter the trade. By 1756 the English whaling fleet had increased to 67 vessels and the Scottish to 16. This increase augured well for the trade, which seemed set for further expansion, in spite of the fact that the whales were being hunted in some of the most treacherous seas in the world. But then came the Seven Years War. The English and French were already in disagreement both in North America and India, and when Frederick the Great of Prussia attacked Austria in 1756 England aligned herself with him, while France supported Austria.

Even though Britain's whaling industry had, at last, been showing signs of coming out of the doldrums, it was not able to satisfy home demands for whale oil, particularly as a means of lighting. These requirements were still largely met by the North American colonies, where more and more ports along the New England coast were entering the whaling trade.

Colonial ships benefited from the Bounty Act of 1749, providing they were 200 tons or more, were built and fitted out in the colonies, and sailed from New England ports. To qualify for the bounty they were obliged to hunt in and around Davis Strait between May and August and bring their cargo to a port in Britain.

The Arctic whaling grounds.
1. Kodiak
2. Bering Strait
3. Bering Sea
4. Kamchatka
5. Sea of Okhotsk
6. Point Barrow
7. Pacific Ocean
8. Lancaster Sound
9. Baffin Bay
10. Melville Bay
11. Baffin Island
12. Davis Strait
13. Disco Island
14. Greenland
15. Greenland Sea
16. Jan Mayen Island
17. Spitsbergen
18. Bear Island
19. Barents Sea
20. North Cape
21. Iceland
22. Faeroe Islands
23. Orkney
24. North Sea
25. Hull
26. London
27. Bay of Biscay
28. Atlantic Ocean
29. Azores
30. Arkhangel

Hunting the whale in Davis Strait.

Author's collection

New England whalers had a set-back, however, when in 1755 the British placed an embargo on the Grand Banks because of the dangers resulting from the war with France. Nantucketers appealed against the ban, which was still in force in 1757, but although it was lifted their whaling activities were restricted by the risks of operating in that area. While enemy navies and privateers sailed the seas no ship was safe. Whalers from Britain as well as from the American colonies suffered and ships bringing cargoes to England fell prey to enemy ships.

Following Wolfe's capture of Quebec in 1759, the Strait of Belle Isle was opened to colonial whalers in 1761. All whale products brought to England from the colonies were subject to a duty imposed by the British to assist their own whaling industry. In addition, a bounty was granted to British whalers in which the colonials did not share. Shortly afterwards the colonials were forbidden to send their exported whale products to any other market than England, and were thus compelled to pay duty on them. These burdensome regulations were among many covering all aspects of trade which affected the livelihood of the American settlers and led to unrest.

The whaling trade was affected by the war in other ways: in Britain the largest and best ships were taken into service as transport vessels, and in the autumn of 1757 all whaleboats belonging to London merchants engaged in the trade were purchased by the government for a proposed expedition against France. Because whalemen were good and experienced sailors, press-gangs were fairly active among them. Naturally this brought protests from the whalers, who resisted in any way possible – from fighting to appealing to magistrates and even Parliament.

While the expansion of the British whaling trade was held up by wartime restrictions, the Dutch, who were not engaged in the conflict, still pursued the whale with success. British whalers ran the risk of being attacked by enemy vessels and, although the Royal Navy was virtually in command of the seas by 1760, the danger was still there until the war ended in 1763. Because of the risks and the poor whaling, both Hull and Whitby abandoned the trade in 1762, but four years later Hull was figuring in the industry again when Captain Standidge (later Sir Samuel Standidge) bought and fitted out an old whaling ship; this supplemented its catch of one whale with 400 seals, thus making the expedition profitable, From then on, seal products are mentioned in many ship's cargoes brought back from the Greenland seas, especially when the number of whales caught was below average.

In New England the whaling industry was spreading to many ports, including, in 1755, New Bedford, a place which was to become the leading whaling port in the country. Joseph Russell, who founded the town, was the first to engage in the trade, which led to great prosperity. Ten years later the four sloops, sailing out of New Bedford, were voyaging, in the summer months, as far south as Virginia. As whales were still plentiful near the land, the blubber was brought back to be processed in try-houses close to the shore. Longer voyages and greater cargoes brought bigger ships into the whaling trade, and by about 1770 New Bedford men, following the example of the Nantucketers, were sailing across the Atlantic.

Sectional drawing of a whaler of about 1720.

Musée de la Marine, Paris

Dutch whalers. This scene on a Delft Tile was executed by Jan Schenk about 1720.

Author's collection

An American whaler lowering her boats.

Author's collection

Up to the outbreak of the American War of Independence in 1775, the story of whaling in the colonies is one of expansion and seizure of opportunity. In 1774, the American whaling fleet comprised 360 vessels totalling 33,000 tons and employing 4700 men. Of these ships, 300 sailed out of Massachusetts ports (half of them from Nantucket), and brought back 30,000 barrels of oil which sold for £167,000. That year American whalers crossed the equator and found an abundance of whales on what became known as the Brazil Banks. Whalers were also sailing regularly to the Cape Verde Islands, the Gulf of Mexico, the West Indies, the Caribbean and by 1775 as far south as the Falkland Islands.

Such was the growth of the American whaling industry that it came in for comment from Edmund Burke in his important speech on Conciliation with the Colonies, in Parliament, on 22 March 1755: '. . . Look at the manner in which the people of New England have of late carried on the whale fishery. Whilst we follow them amongst the tumbling mountains of ice, and behold them penetrating into the deepest frozen recesses of Hudson's Bay and Davis's Straits . . . Nor is the equinoctial heat more discouraging to them, than the accumulated winter of both poles. We know that whilst some of them draw the line and strike the harpoon on the coast of Africa, others run the longitude, and pursue their gigantic game along the coast of Brazil. No sea but what is vexed by their fisheries. No climate that is not witness to their toils. Neither the perseverence of Holland, nor the activity of France, nor the dexterous and firm sagacity of English enterprise, ever carried this most perilous mode of hardy industry to the extent to which it has been pushed by this recent people; a people who are still, as it were, but in the gristle, and not yet hardened into the bone of manhood.'

The whaling activities of Burke's own countrymen were also on the increase. Regulations drawn up in 1771 gave a ship exemption from paying duty and enabled it to claim a bounty provided that the captain and one third of its crew were British; that it was fitted out correctly for whaling in the Greenland Sea, Davis Strait or adjacent seas; that whaling was carried out within the proper season; and that the ship returned to the port from which it sailed. A vessel of 200 tons had to have a crew of 30 men, including the master and surgeon; an extra six men had to be taken on for every 50 tons over 200 tons. The ship had to engage one apprentice for every 50 tons, in order to encourage young men into the trade. The size of the ship also determined the number of whaleboats to be carried: four boats on a 200-ton vessel, and an extra boat on one between 200 and 400 tons. If these regulations were complied with, the ship qualified for a bounty of 50*s* per ton for every voyage made during the first five years, 30*s* in the second five years and 20*s* in the third five years, with the date of expiry fixed at 25 December 1786. These bounties were also applicable to any British-American whaling vessel not more than two years old, which left its American port before 1 May and, after hunting in the Greenland seas, brought its produce to Britain.

However, in North America, grievances over taxation had set the colonies on a collision course with the British government. Colonial anger was particularly aroused by the tax on tea and the East India Company's monopoly on tea imports into the colonies. Three ships, the *Dartmouth,* the *Eleanor* and the *Beaver*, owned by William Rotch and his brother of Nantucket, delivered a cargo of whale oil to London in

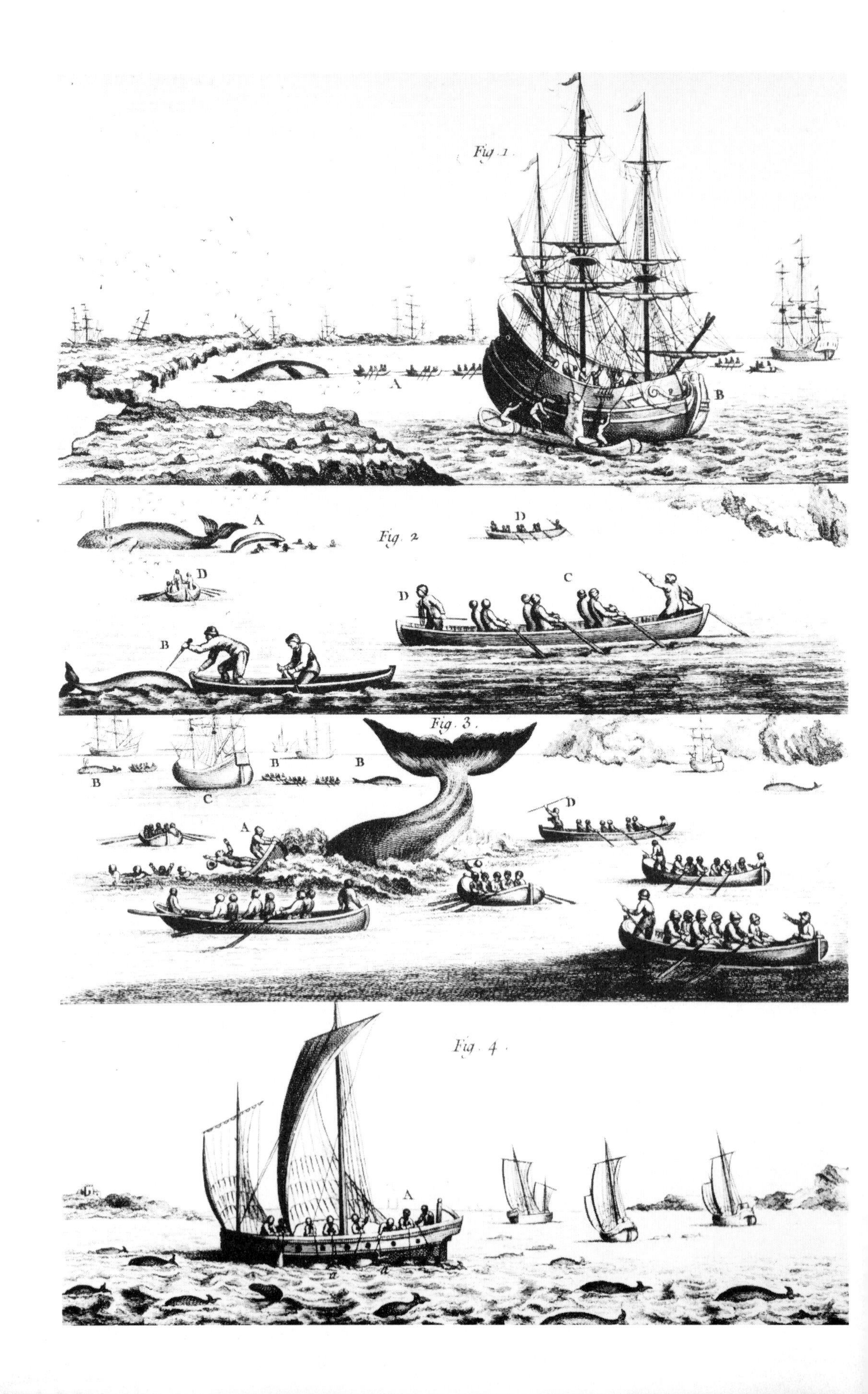
Fig. 1.
A
B
Fig. 2
A
D
D
D
C
B
Fig. 3.
B
B
B
C
A
D
Fig. 4.
A

Whaling scenes from a treatise by Duhamel du Monceau, 1782.
Fig 1: A whale is towed, with the help of a hook fixed in its mouth, to the ship, where large pieces of blubber are hoisted on board.
Fig 2: A boat has been overturned by a whale. A man pricks a whale with a bayonet to make sure it is dead. Other boats stand by to give assistance.
Fig 3: The whale attacked by a number of boats overturns one of them, while in the background whales are being towed to the waiting ships.
Fig 4: Fishing from the side of the boat with lances and harpoons.

Musée de la Marine, Paris

1772. The ships were then chartered by the East India Company for the transportation of tea to New England. They docked in Boston, where on 11 December 1773 they were boarded by colonists who threw the tea overboard in what has become known as the Boston Tea Party.

War soon followed and British whaleships were commandeered to transport soldiers and supplies for the campaign in North America. Those whalers which were able to follow their trade had difficulty in raising crews, as more and more men were pressed into naval service, and they were always in danger of attack by enemy ships.

In 1776 the bounty was reduced to 30*s* and there was talk of stopping it altogether, which would have meant the death of the whaling industry. The mayor and burgesses of Hull, in anticipation of the end of the war and the revival of a profitable trade, petitioned Parliament not to take such a step. Their plea was successful, but in 1782 they had to approach the government again to ask for a bounty of 40*s* to enable the industry to pay its way. The government agreed to increase the bounty and also allowed the whalers to call at the Shetland Islands to make up their complement of crews. This practice was followed by the whalers from the east coast ports throughout their whaling days, the islanders being expert seamen and proving a valuable addition to the ships' crews.

'Whalers in the Arctic', painted by John Ward of Hull.

By courtesy of Kingston-upon-Hull Museums

'The Northern whale fishery'. This whaling scene was painted in 1778 by Hendrik Kobell who was born in Rotterdam in 1751. He trained in Amsterdam and London and died in 1799.

Author's collection

A whaler in the Arctic.

By courtesy of Kingston-upon-Hull Museums

The American War of Independence was a tragedy for New England's whaling industry. At a point where expansion was forging ahead, the fleet fell prey to the British, who swept it from the sea, by capturing, sinking or burning nearly all the ships. The whaling towns, too, suffered from attack: New Bedford, which had between 40 and 50 whalers at the outbreak of war, lost thirty-four of these in one raid when Sir Henry Clinton's expedition attacked the port in 1778. It was the same story all along the coast, with every whaling fleet almost destroyed.

The people of Nantucket, whose livelihood depended on the sea, attempted to remain neutral, with the result that their ships faced attack from both sides. Because whaling was so important to them they were forced to pursue the trade in spite of the risks involved. As ship after ship was destroyed, particularly by the British, the plight of the Nantucketers became desperate. They petitioned both sides for permission to carry on whaling unmolested. The British granted 25 permits in 1780, but similar permits from the American Congress did not come until 1783, too late to be of practical use. Nantucket lost 149 of her 150 whale ships and of the 800 families on the island at the end of hostilities 202 consisted of widows and children.

Six months after the outbreak of hostilities Francis Rotch organised sixteen whaleships to sail from Nantucket and New Bedford as far south as the Falkland Islands. Instead of returning to New England the ships were to take their oil direct to London. Rotch and one of his partners were to go to London from where they would continue to fit out ships for the Southern Fishery, so called to distinguish it from the

The whaling vessel *Swan* of Hull.

By courtesy of the Wardens and Brethren of the Corporation of Hull Trinity House

The whale and its uses – an educational illustration of the nineteenth century showing The Northern whale-fishery (centre), the sperm whale-fishery (top), the various uses of the whale, as illuminant (top left, top right), for food (left centre), whalebone for umbrellas (bottom left), for manure (bottom centre left), for oil (bottom centre right) and for its spermaceti and ambergris used in various branches of commerce (bottom right).

Mansell Collection

Northern (Arctic) Fishery. If his plans had come to fruition the British would have had a ready-made Southern Fishery and could easily have developed it in the way the Americans did much later. However, Rotch was unfortunate. The Royal Navy captured several whalers on their way to the Falkland Islands, and one with a cargo of oil was forced to surrender to John Paul Jones of the US Navy. Although Rotch received the release of the captured ships he did not have sufficient to make the venture he had planned a viable proposition as the war escalated.

He had, however, set an idea going in the minds of several London merchants closely associated with the whaling trade and with the Nantucketers. Notable among the former was Samuel Enderby, who sent British ships, manned by Nantucket whalemen skilled in hunting the sperm whale, to the Southern Fishery. His ships were successful and other merchants followed his example.

With American whaling virtually non-existent, with the Dutch industry declining, and with the nucleus of a whaling fleet for northern waters and whalers sailing to the Southern Whale Fishery, Britain had a unique opportunity to expand and seize control of the world whaling industry.

Sperm Whale Fishery
For Light ... as a Guide to Mariners
Cutting off the Blubber
Published by THOMAS VARTY, 31 Strand, London
Agriculture ... For Manure
Manufacture ...Oil Works
FINE
AMBERGRIS
&
SPERMACETI
Commerce ... Spermaceti, Ambergris

5 The British Boom 1783-1812

On 3 February 1783 the whaleship *Bedford*, belonging to William Rotch of Nantucket, sailed into London with a good cargo of oil, the first vessel to fly the flag of the new nation in British waters. After some controversy the Customs allowed the oil to be unloaded. The scarcity of whale products after the war ensured a good price for the cargo as it did for that brought in by the brig *Industry* on 29 March. New England merchants, sensing a possible boom, returned to their former trade with high hopes. Soon whalers were sailing from Nantucket, Boston, Wellfleet, Braintree, Providence, New London, Sag Harbour, Hingham, Plymouth, Bristol and Newburyport.

Faced with the prospect of a flood of whale oil from a nation to whom she now felt no obligation, Britain, determined to protect her own whaling fleet, imposed a duty of £18 per ton on imported whale products. The Americans were priced out of their only overseas market, while the British, with their expanding fleet, were becoming less and less dependent on such imports.

A bounty offered by the Massachusetts Legislative in 1783 to ships owned and crewed by residents of the state and bringing their products to a Massachusetts port provided a stimulus. Unfortunately the demand for oil had decreased, due partly to people getting accustomed, during the war, to using tallow candles which, though not as good, were cheaper than spermaceti candles. With an abundance of oil and no foreign outlet, the price dropped to a point where it ceased to be economical to produce it, and some ports, seeing few prospects in the whaling trade, began to abandon it.

But whaling was the only livelihood the Nantucketers knew. It was necessary to their survival so it was natural that, when all hopes of recovery of the whaling industry on their island seemed blocked, they should seek to follow their trade elsewhere. Some came to London to serve in the fleet sailing to the Southern Whale Fishery while others examined different possibilities. Forty families, by an agreement which gave them British citizenship thereby enabling their whale products to

Whalers entering Whitby. Whitby, on the Yorkshire coast of England, was one of the leading whaling ports in the late eighteenth and early nineteenth centuries. This ship, the *Phoenix*, was employed in the Arctic from 1826 until 1833.

By courtesy of the Whitby Literary and Philosophical Society

Whaling Ports of the British Isles
1. Lerwick
2. Stromness
3. Kirkwall
4. Banff
5. Peterhead
6. Aberdeen
7. Montrose
8. Dundee
9. Greenock
10. Glasgow
11. Grangemouth
12. Bo'ness
13. Queensferry
14. Leith
15. Dunbar
16. Berwick
17. Kirkcaldy
18. Whitehaven
19. Newcastle
20. Sunderland
21. Stockton
22. Whitby
23. Scarborough
24. Hull
25. Grimsby
26. Liverpool
27. Lynn
28. Yarmouth
29. Ipswich
30. London
31. Bristol
32. Exeter

enter Britain duty-free, left Nantucket and settled in Nova Scotia. This venture was highly successful and during its seven years of existence excellent returns were made on whales chiefly taken in the Gulf of Guinea and the Brazil Grounds.

In 1785 William Rotch came to England from Nantucket with a proposal that £20,000 should be given to 100 families who would come to England and use their experience for the benefit of the British whaling trade. The government dithered, offering £87 10s per family, so Rotch went to France where his negotiations were successful. He gained special terms and privileges for the Nantucketers who were to settle at Dunkirk and, mindful of the needs of his native island, he also gained the concession to bring into Dunkirk 250 tons of oil duty-free. Nantucket whaling was kept alive: it could be nurtured, and if the conditions were right it could thrive. Now the Nantucket and Dunkirk fleets could set about trying to catch up and challenge the British role in the Southern Whale Fishery.

But Britain dominated that sphere of whaling for a considerable time. The value of whale products brought from this branch of the trade in 1789–90 was £84,493 from 51 ships, whereas that from three times the number whaling in the Arctic was £66,662, such were the different values for the different oils. Sailings to the vicinity of the Falkland Islands and the Patagonian coast were a regular feature and thoughts of the possibilities off the west coast of South America were being entertained when the *Emilia*, a whaler belonging to Samuel Enderby & Sons, sailed from London on 7 August 1788.

In January 1789 the captain, following orders, took his vessel round Cape Horn into the Pacific. On 5 March the first whale to be killed in the Pacific was credited to the *Emilia*'s first mate, Archelus Hammond, one of several Nantucket men serving in the crew of twenty-one. The *Emilia* reached London in January 1790 with 147½ tons of sperm oil and a report of numerous whales off the coast of Peru. The British lost no time in capitalising on this pioneering voyage and in the period 1790–93 had 23 whalers in the Pacific. The Americans soon followed and in 1791 the *Beaver* of Nantucket was the first American whaler to round Cape Horn into the Pacific, being followed by the *Washington, Hector* and *Rebecca*, all from the same port.

So the Pacific had been brought into whaling history; pioneered by the British, it would be developed by the Americans on a vast scale, taking them through their golden age of whaling.

'The *Molly* and the *Friends*'. These two ships were built in America, the *Molly* of 290 tons in 1759, the *Friends* of 256 tons in 1778. They became whalers out of Hull, England, the *Molly* making her first voyage for whales in 1775 and the *Friends* in 1786. The *Friends* was lost while sealing in 1790.

By courtesy of the Kingston-upon-Hull Museums

Whalebone-scrapers at Whitby, 1813. Because of its great strength and lightness whalebone was wanted for stays, umbrellas, carriage springs, upholstery, fishing rods, knife handles, frames for hats, sieves, nets and brushes. Picture from *Costumes of Yorkshire*, published 1814.

By courtesy of the Whitby Literary and Philosophical Society

Mindful of the long voyages involved in whaling in the Pacific, British merchants realised that their vessels should have suitable places to refit and replenish supplies independent of the Spanish ports on the coast of South America. They requested that a naval vessel be sent to find such places and on 2 January 1793 HMS *Rattler*, captained by James Colnett, who had experience of the Pacific and had served on Captain Cook's second voyage, left England for this purpose. The *Rattler* was also fitted out for whaling, but in this respect her voyage was unsuccessful, Colnett having had no whaling experience since he was first and foremost an explorer, in which capacity he did some valuable work.

In the meantime British merchants, interested in the whaling trade, had also seen an opportunity on the other side of the Pacific. With the loss of the American colonies Britain viewed Australia as a settlement for convicts. In the first fleet which sailed on 13 May 1787 and arrived at Botany Bay in January 1788 were two whalers with licence to go whaling after completing the voyage to Australia. It is not known if they did any whaling but they did report whales in the Indian Ocean.

The owners of British whaleships appreciated the advantages of transporting convicts and supplies to Australia on their outward voyage. In the third fleet, which sailed in 1791, were five whaleships. Whales were sighted off the coast of Tasmania and, as soon as they got clearance from Sydney, the *Britannic* and the *William and Ann* left on 24 October 1791 to hunt whales, the first whalers to do so from Sydney.

The stage was set for further developments and more and more whalers made the voyage from England. They often served in the role of merchantmen once they had reached Sydney, because of the precarious position of the new colony's supplies, and they played no small part in the continued existence of the settlement. They discovered new whaling grounds and explored areas not on the schedules of official exploring expeditions. The whalers fulfilled an important role in the development of that part of the world and it was they who finally broke the monopoly held by the East India Company in these seas when an Act of Parliament was passed giving them the right to carry on their trade without a licence from the East India Company or the South Sea Company. Sydney became a thriving settlement, important as a port of call, especially for the whaleships which pursued their trade in the Pacific. From here the New Zealand whaling grounds were opened up and by 1801 whalers of all interested nations were using the bays and inlets on the New Zealand coast, especially the Bay of Islands, as bases in their hunt for the sperm whale. In 1803 the whaleship *Albion* which had come from England was chartered to carry settlers and convicts as well as supplies to Tasmania. Three sperm whales were taken en route, an unusual occurrence when convicts were on board. The settlers founded what was to become Hobart and saw an abundance of whales in the Derwent River. Almost immediately William Collins initiated bay whaling. His idea was to combine sperm whaling in the summer with bay whaling in the winter and spring, but right whales were so plentiful that whaling from Tasmania followed the pattern seen in other parts of the world – a concentration of offshore whaling first.

Through her colonies Britain was finding another source of whale products. As Australia developed her own whaling industry it made sense to import this oil into Britain rather than send whalers on the long voyage to the Pacific. This situation can be put forward as one of the main reasons why Britain did not continue to expand her Southern Whale Fishery and develop it in the same way as the Americans did.

Meanwhile the success of the Nantucketers who had settled at Dunkirk was causing the British some concern. To counteract this they persuaded the Nantucket colony in Nova Scotia to move to Milford Haven and run their whaling industry from there. The move took place in 1792 and 1793, but the success it could have achieved was hampered by opposition in certain government quarters and from London whaling merchants who saw it as a threat to their own successful trading. When France declared war on Britain on 1 February 1793 the people of Dunkirk found themselves in a precarious position. William Rotch left Dunkirk and, after a brief stay in England, returned to America. Seeing the developments in New Bedford he transferred his firm of William Rotch & Sons there. Nantucket had regained some of its former whaling strength and with the fillip given to the trade by the growth of New Bedford the American whaling industry was moving on to a strong base for future prosperity.

The Yorkshire port of Whitby, England, entered the whaling trade in 1753, abandoned it in 1762, re-entered it in 1767, became one of Britain's leading whaling ports and finally left the trade in 1837. During the port's engagement in whaling 58 ships left on a total of 577 voyages. The longest serving vessel was the *Volunteer* which sailed 54 times to the Arctic for whales (1772–1835). She was built in Whitby in 1756, served as a trader and, after her career as a whaler, reverted to the merchant trade sailing from Hull. She was 84 years old when she left there for Sierra Leone in 1841.

Author

The whale's jaw-bone erected on Whitby's west cliff stands as a reminder of the important days of Whitby's whale trade. Although this particular jaw-bone does not survive from those days (it was presented to Whitby by the Norwegians in 1960) similar erections were once a common sight throughout north-east Yorkshire where the jaw-bone of the Greenland whale was used as archways over entrances to farms and houses.

Author

Australia
1. Fremantle
2. Geographe Bay
3. Augusta
4. Albany
5. King George Sound
6. Doubtful Island Bay
7. Victor Harbour
8. Encounter Bay
9. Port Elliot
10. Portland Bay
11. Hobart
12. Two Fold Bay
13. Sydney
14. Mosman's Bay

Even while Britain's Southern Whale Fishery was in a strong position the development of her Arctic whaling was growing, and by the late eighteenth century she was the foremost whaling nation in the world.

The number of vessels sailing to the Arctic in the 1780s grew steadily and reached 255 in 1788, London sending 91 ships, Hull 36, Liverpool 21, Whitby and Newcastle 20 each, Yarmouth and Sunderland 8, Leith and Lynn 6, Ipswich and Dunbar 5, Bo'ness, Aberdeen and Glasgow 4, Montrose and Dundee 3, Greenock, Whitehaven, Stockton and Exeter 2, and Scarborough, Grangemouth and Queensferry 1 each. It is an interesting list, including as it does places which never participated in the trade to a great degree, while Dundee, which appears well down the list, was to become a prominent whaling port. With 21 ships, Liverpool achieved its pinnacle of importance; however, Peterhead, a port which later was to provide some of the most illustrious names in whaling history, is not mentioned. London was still in the lead but before the end of the eighteenth century it had been overtaken by Hull.

Among the ships which left Hull in 1784 was its most famous whaler, the *Truelove*, built in 1764 in Philadelphia for the merchant trade. During the American War of Independence she was used by the colonists as a privateer. Captured by a British cruiser, she was sold by the government to Hull merchants who used her for the shipment of wine from Oporto in Portugal. She was then converted to a whaler, continuing in this capacity until withdrawn after the voyage of 1868, when whaling in the Greenland seas was no longer regularly profitable.

On board the *Henrietta* out of Whitby in 1785 was William Scoresby, serving as an ordinary seaman. He was to gain rapid promotion and become one of Britain's best-known whaling captains. He and his son William, who also commanded vessels out of Whitby, were largely responsible for the good catches which at one time gave their home port a leading position in the industry.

The Bounty Act of 1786 reduced the bounty to 30*s* for ships of 150 tons and over for a period of five years, after which no vessel over 300 tons would receive a bounty unless it had been used for whaling before this act was passed. Vessels had to fulfil certain other conditions and had to sail by 10 April and remain in the whaling areas until 10 August unless such circumstances as the risk of being caught in the ice prevented this.

Ice was not the only hazard encountered by the whalemen. As they were expert sailors much in demand by the navy, press-gangs were active among them. In 1790, the authorities went so far as to post the *Racehorse* at Hull for the purpose of impressment, much to the annoyance of the whalemen and the citizens of the town. When the whaler *Eggington* docked on her return, her crew came ashore armed with harpoons and lances, ready to resist the press-gang from the *Racehorse*, while other whalers put men ashore along the Yorkshire coast to avoid impressment.

New Zealand

1. Doubtless Bay
2. Bay of Islands
3. Auckland
4. Wellington
5. Cook Strait
6. Picton
7. Blenheim
8. Christchurch
9. Banks Peninsular
10. Otago Peninsular
11. Dunedin
12. Campbell Town
13. Foveaux Strait
14. Preservation Inlet
15. Cloudy Bay

0 50 100 150 200 miles

N

'Northern whale fishery', an engraving by E Duncan after a painting by W J Huggins. From left to right the vessels are the *Margaret* of London, the *Eliza Swan* of Montrose, the *Harmony* of Hull and the *Industry* of London.

By courtesy of Kingston-upon-Hull City Library

War with France prevented an expansion of the trade since whaleships were commandeered as transports, and those that did sail to the Arctic ran the risk of attack by enemy vessels. The Arctic fleet set out in March or early April and returned in August or September. The ships sailed north along the east coast, calling at the Orkneys and Shetlands where they completed their crews. Apart from being excellent seamen, the islanders were generally quiet by nature and exerted a steadying influence on the rest of the crew. The people of Stromness and Lerwick welcomed the whalers in more ways than one, offering all manner of entertainment to men who would have no contact with civilisation for about four months.

'The *Truelove*' by W J Ward, 1801. Built in Philadelphia, USA in 1764 and captured by a British cruiser during the American War of Independence, *Truelove* was purchased from the government in 1780 and engaged in the wine trade between Oporto and Hull until converted into a whaler for the 1784 season. Her last voyage as a whaler was in 1868.

By courtesy of Kingston-upon-Hull Museums

A certain glamour attached itself to Greenland whalers, as these men were known, for they combined the qualities of a hunter and an explorer, qualities needed by men who faced the rigours of the northern seas. No matter what their background nor how uncouth their manner, something of this glamour rubbed off on them. There was also a gambling instinct in those who sailed with the whaling ships, for a man's rewards over and above his basic wage depended on the success of the voyage. If the whaling was good, then on return to port his pay was good, which meant a comfortable winter without hardship for himself and his family. If, on the other hand, the catches were poor or the ship returned clean, they faced a bleak winter. Consequently, crews were attracted more easily to ships whose captains were known to be particularly skilful or 'lucky' on their whaling expeditions.

The life was hard and a man required toughness and courage to cope with the demanding work and survive the dangers which threatened from the sea, ice, weather and, not least, the prey itself. In the fo'c's'le, where very often a man could not stand upright, it was frequently damp and dirty and the only light was provided by candles or lamps; there the men slept in two-tiered bunks fixed along the sides. Conditions were little better for hands housed amidships, where they slept on straw mattresses. Their bedding was often damp, for rough seas could mean flooding. The few possessions a man had were kept in his sea-chest, which was lashed to a ring fixed in the fo'c's'le deck and which served also as a seat on which, with a tin plate balanced on his knees, he ate his food.

'*Isabella* and *Swan* in the Arctic regions' by John Ward. The Hull whaler *Isabella*, 374 tons, made her first voyage to the Arctic in 1786. The *Swan*, also of Hull, was 323 tons and first hunted whales in 1816. John Ward was born in Hull in 1798. He was a house- and ship-painter until 1838 when he became a full time artist with a number of pupils. It is thought that he made two whaling voyages with his father, Abraham Ward, who was also an artist. John died of cholera in Hull in 1849.

By courtesy of the Wardens and Brethren of the Corporation of Hull Trinity House

'The *Munificence*' by R Willoughby, 1805. Built in 1761, the *Munificence* sailed on her first whaling voyage in 1802 from Hull to Davis Strait. Robert Willoughby, 1768–1843, lived in Hull and developed a particularly individual style of painting whaling scenes and vessels.

By courtesy of Kingston-upon-Hull Museums

A hard day's work, much of it in the cold frosty air, stimulated the appetite, but fresh food lasted no longer than a few days out of port; after that it was salt beef and salt pork, often of poor quality, smelling unpleasant and badly cooked. The fresh bread was replaced by hard biscuits, which often developed weevils. Scurvy, caused by the lack of fresh food, was a dreadful hazard, and the risk of contracting it increased when ships were beset by ice and forced to winter in the Arctic.

The whaling ships, which might be as large as 400 tons, were mostly three-masted square-riggers, but there were some topsail schooners and an occasional brig. These vessels were strengthened to withstand the tempestuous seas and crushing ice by a method called 'doubling and trebling'. Two layers of oak planking reinforced the hull, while the bows were trebled with an additional outer layer of greenheart.

Between four and seven whaleboats were carried amidships, but once the whaling area was reached they were slung out on davits with their full equipment, ready for instant action when there was a sighting. Each whaleboat was from 25 to 28ft long and manned by six or seven men, one as a helmsman the rest as oars. The five-oared boat was the one in general use. The harpooner was in command of the boat and he pulled the bow oar until they were 'on fish', that is, in a position to strike.

The Northern whale-fishery.

From 'The Marine Mammals of the Northwestern Coast of North America' by Charles M Scammon

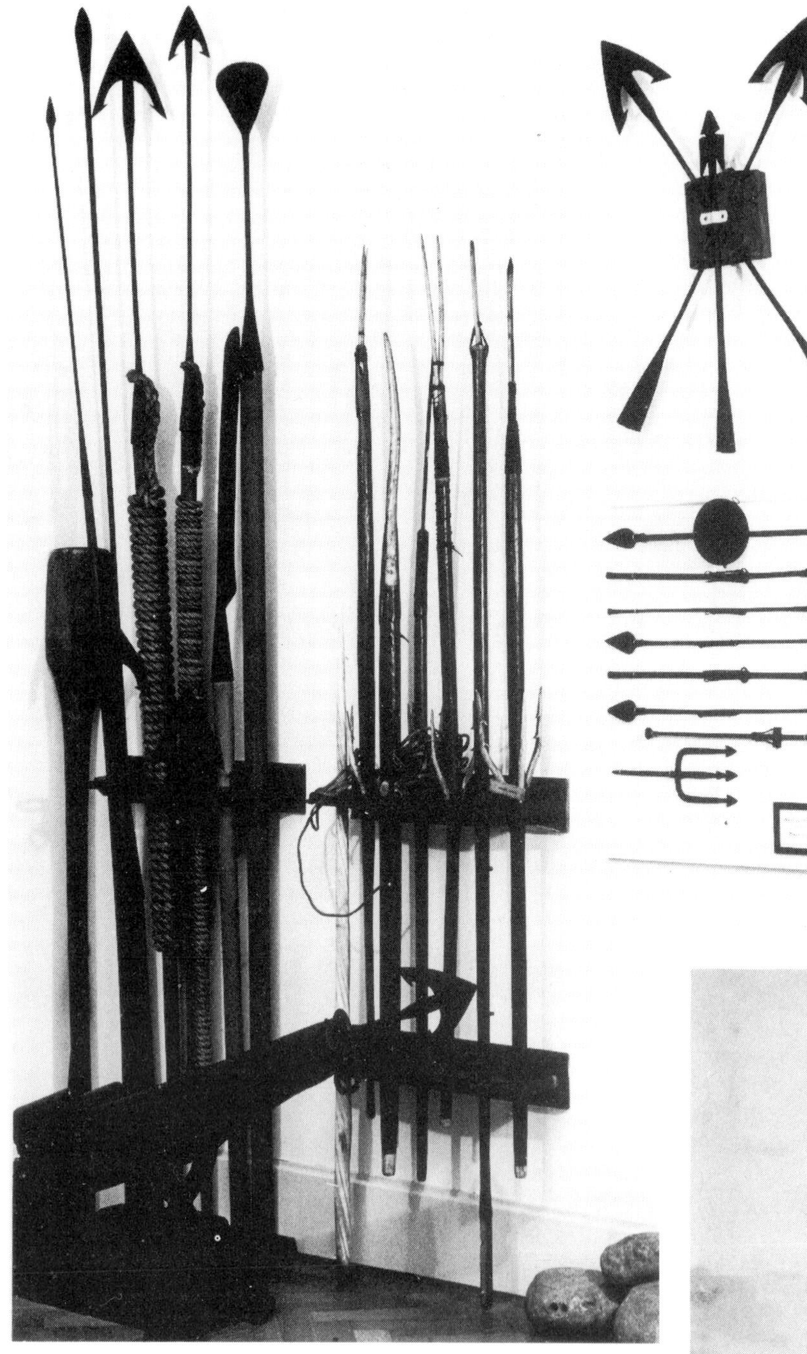

Implements used in the whaling industry. Left to right, they are a snow shovel, lances, harpoons, flensing knife, blubber cutter, various Eskimo harpoons, and harpoons with sealing harpoons below. In front stands an early harpoon gun.

By courtesy of the Whitby Literary and Philosophical Society

'The *Diana* of Hull'. This vessel, one of three Hull ships bearing the name *Diana*, was of 355 tons and made her first voyage in 1785.

By courtesy of the Wardens and Brethren of the Corporation of the Hull Trinity House

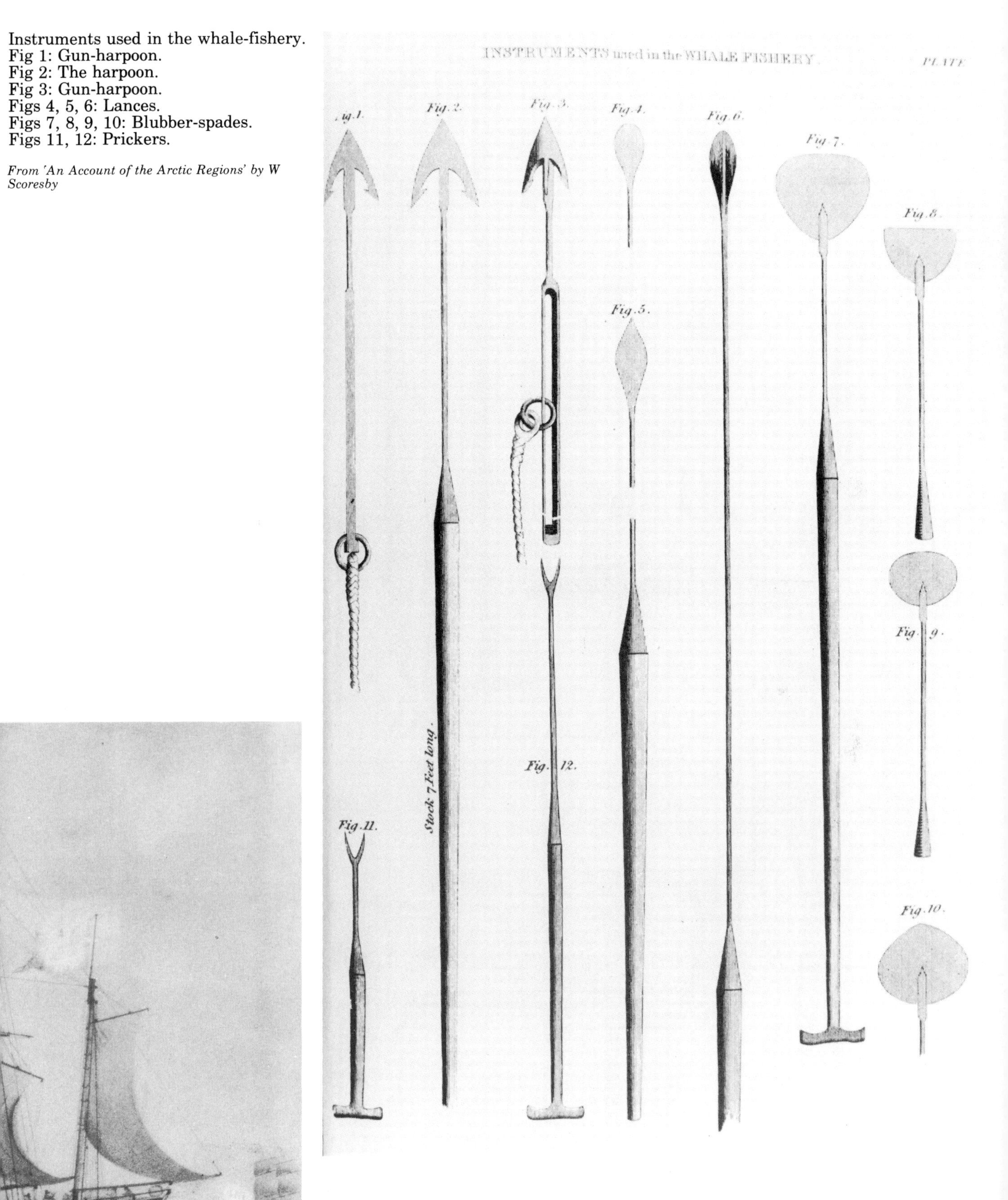

Instruments used in the whale-fishery.
Fig 1: Gun-harpoon.
Fig 2: The harpoon.
Fig 3: Gun-harpoon.
Figs 4, 5, 6: Lances.
Figs 7, 8, 9, 10: Blubber-spades.
Figs 11, 12: Prickers.

From 'An Account of the Arctic Regions' by W Scoresby

Apparatus used in the whale-fishery.
Fig 1: Blubber-knife.
Fig 2: Chopping-knife.
Fig 3: Strand-knife.
Fig 4: Tail-knife. For perforating the fins or tail of a dead whale.
Fig 5: Bone-geer. For handling the whalebone (baleen).
Fig 6: Bone-wedge. For splitting the whalebone (baleen).
Fig 7: Mik, or rest for the harpoon.
Fig 8: Third-hand. Used in flensing.
Fig 9: Pick-haak.
Fig 10: Closh. For holding blubber while skinning.
Fig 11: Grapnel.
Fig 12: Ice-grapnel. Used in warping.
Fig 13: Krenging-hook.
Fig 14: Krenging-knife. Tools of Krengers, who removed muscle and fat, from blubber.
Fig 15: Spurs. For use in standing on dead whale, during flensing.
Fig 16: Axe. For cutting the lines when necessary.
Fig 17: Snatch-block.

From 'An Account of the Arctic Regions' by W Scoresby

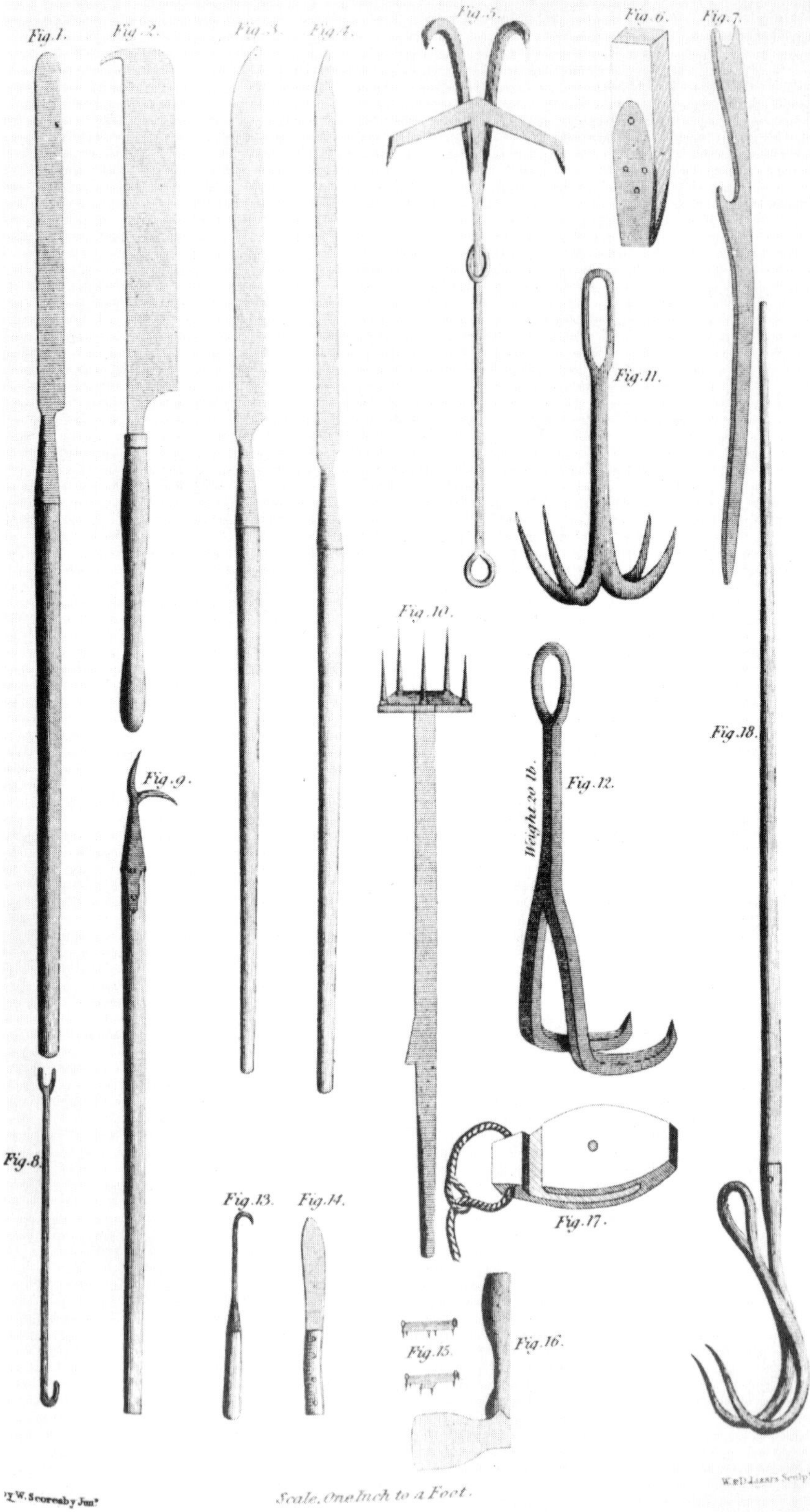

Apparatus used in the Northern whale-fishery.
Fig 1: Ice-axe.
Figs 2, 3: Ice-anchors.
Fig 4: Bay ice-anchor.
Fig 5: Blubber-pump. For removing ballast water from blubber casks.
Fig 6: Bone hand-spike. Used for dislodging whalebone (baleen).
Fig 7: Ice-saw. Length 14ft. When used for cutting ice to release the ship, two handles were put through the rings at the top so that 12 to 16 men could work it together.
Fig 8: Ice-saw. Used for thinner ice.

From 'An Account of the Arctic Regions' by W Scoresby

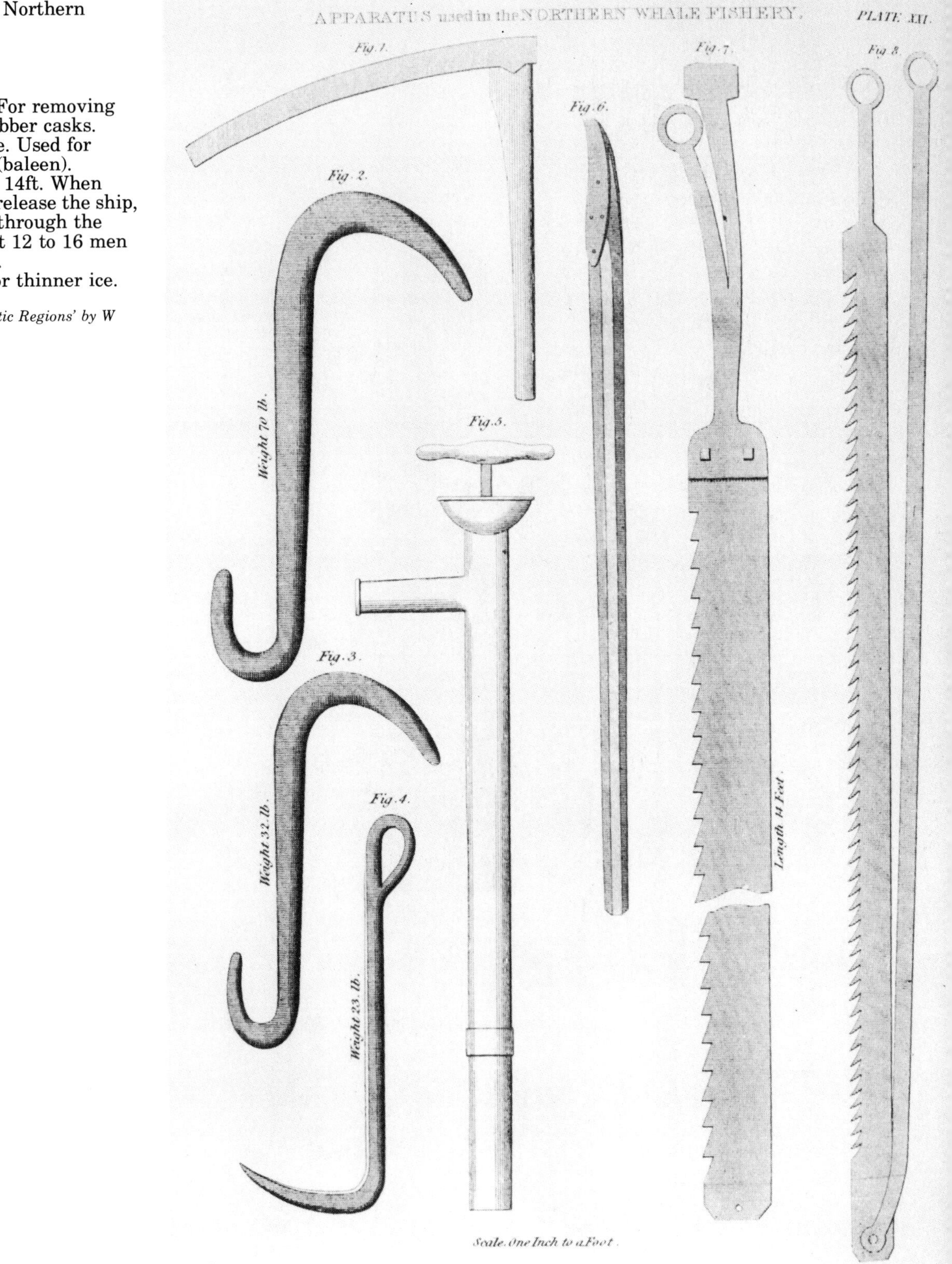

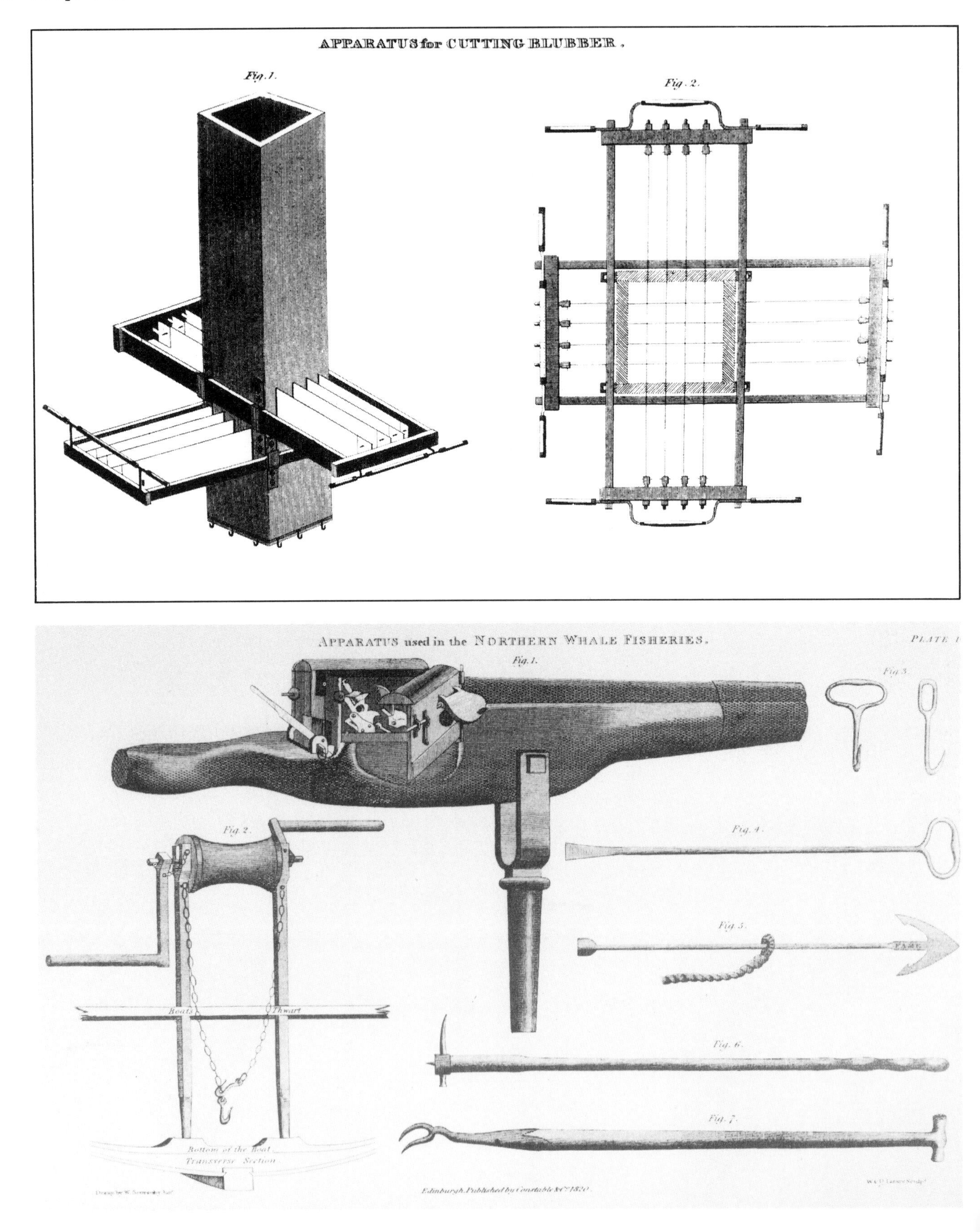
APPARATUS for CUTTING BLUBBER.
Fig. 1.
Fig. 2.
APPARATUS used in the NORTHERN WHALE FISHERIES.
Fig. 1.
Fig. 2.
Fig. 3.
Fig. 4.
Fig. 5.
Fig. 6.
Fig. 7.
Boats
Thwart
Bottom of the Boat
Transverse Section
Edinburgh. Published by Constable & Co 1820.

This apparatus for cutting blubber was introduced shortly before 1820 in the Arctic trade. The square tube of wood was 5 or 6ft in length, and about 16 to 18in square. An iron frame with four cutting knives, edges upwards, was fixed on rollers near the bottom of the tube. Below this was another similar frame at right angles. The blubber was inserted in the top tube, and fell on to the edge of the knives, which were put into rapid horizontal motion thus cutting through the blubber. Fig 1 shows the instrument; fig 2 is a horizontal section.

From 'An Account of the Arctic Region' by W Scoresby

Apparatus used in the Northern whale-fishery.
Fig 1: Harpoon-gun.
Fig 2: Boat's winch. Used in whaleboats for heaving in the line.
Fig 3: Hand-hooks. For handling blubber.
Fig 4: Ice-drill. For setting anchor into ice.
Fig 5: A gun-harpoon.
Fig 6: Seal club.
Fig 7: King's fork. For moving pieces of blubber.

From 'An Account of the Arctic Regions' by W Scoresby

Once the dead whale had been brought alongside the ship it was secured to it by its head and tail, and held up out of the water by means of a rope threaded through one of the flippers. While one or two of the whaleboats lay alongside to lend a hand, the harpooners, with spikes strapped to the bottom of their boots, to help them keep their footing, cut the blubber from the whale.

Using 'spades' with long handles and sharp blades, the initial cuts were made in such a way that the end of each long strip of blubber could be raised sufficiently for a hole to be made and a rope passed through, which was prevented from slipping back by a large wooden toggle. The other end of the rope passed over a heavy tackle secured to the head of the mainmast and back to the deck of the ship, where the crew heaved upon it, raising the strip of blubber, while the harpooners deftly eased it off the whale by making quick decisive cuts with the spades. When a suitable length had been peeled off, it was cut, and the blanket piece, as it was called, was heaved on deck. The procedure was repeated until all the blubber had been stripped from the body. Meanwhile, the blanket pieces were cut up and piled into heaps with long-handled, two-pronged implements known as 'king's forks'.

The heaped-up pieces were then prepared for the casks waiting amidships. The skin had to be removed from the outside, as did any fleshy or fibrous matter from the inside. If these parts were not removed properly, the blubber would be spoiled, and decomposition would result in a build-up of gas which would cause the casks to burst.

When the blubber had been prepared it was chopped into small pieces and passed down a canvas pipe to the waiting barrels, which were filled, bunged and stowed under the supervision of the skee-man, who was responsible for the operations conducted in the hold. The whalebone from either side of the upper jaw was cut out as one piece and, once all this and the blubber had been removed, the carcase or 'krang' was cast adrift.

It was evident in the first years of the nineteenth century that the whale in the Greenland Sea was being gradually exterminated. In his efforts to fill all his casks the hunter was prepared to kill young whales before they reached maturity, rather than make the longer voyage to the more dangerous areas of Davis Strait.

In 1806 French frigates were particularly menacing to the whalers who also had to contend with an ice barrier much further south than usual. Some captains risked battling through it to open sea beyond and were rewarded with a good catch. One of these was William Scoresby who, with his son as chief mate, was in command of the *Resolution* of Whitby. He took the ship through another barrier and came within 510 miles of the North Pole – closer than any man had been. This was a record which stood until beaten by the British explorer William Edward Parry in 1827.

Scoresby was born and bred on the southern edge of the North York Moors, but a visit to the port of Whitby, when he was young, fired in him an ambition which was fulfilled when he was nineteen. In 1785 he made his first voyage on a Greenland whaler and in 1791 sailed out of Whitby as captain of the *Henrietta*. This was rapid promotion, but Scoresby was no run-of-the-mill whaler. He was responsible for many innovations in the whaling trade and invented the now-familiar round topgallant crow's nest, the first being constructed in May 1807. Prior to

Captain William Scoresby Senior was born in 1760 at a farm near Cropton, a tiny village near Pickering in North Yorkshire. He first went to sea in 1780 and in 1785 joined a Whitby whaler. He became a captain in 1791. Before his retirement in 1823 he made 30 voyages in 6 different ships, taking 533 whales. He died in 1829.

By courtesy of the Whitby Literary and Philosophical Society

An enlargement of part of the log of the *Resolution*'s voyage of 1804 shows a method of recording the killing of a whale by drawing the whale's tail under the date with the names of the men who made the kill. The *Resolution*, under Captain Scoresby Senior, made eight voyages to the Arctic and his son took command of this vessel for a further two voyages. She was built by Fishburn and Broderick, at Whitby, for a first voyage made in 1803. In 1829 she was sold to Peterhead.

By courtesy of the Whitby Literary and Philosophical Society

William Scoresby Junior, the son of the famous Whitby whaling captain of the same name, was born in 1789, and first went to the Arctic at the age of ten in his father's vessel the *Dundee*. In 1803 he was apprenticed, under his father, on the *Resolution*, became mate in 1806 and took command on his 21st birthday in 1810. Not only a successful whaling captain, he was also an explorer, scientist and scholar. He wrote a classic of Arctic literature, *An Account of the Arctic Regions*, which was published in 1820. He gave up his whaling life in 1823, and became a minister of the Church of England, keeping up his scientific studies and writings until his death in 1857.

Author's collection

this invention the lookout was positioned on the main topgallant or top-gallant masthead with only a canvas screen for protection and an inadequate seat, the instability of which led to accidents if the lookout became drowsy. A man in such a position in Arctic temperatures might reasonably take less interest in watching for whales and in navigation than in his own survival. Scoresby's invention offered much more protection, since, being enclosed, there was no danger of the lookout falling. This, coupled with a certain degree of comfort, enabled him to concentrate on his important job.

Captain Scoresby's son, William, also became a famous whaling captain. A well-educated young man, he made many scientific observations which were advanced for the time. These and details of the whaling trade he published in a remarkable work, *An Account of the Arctic Regions*, in 1820.

The 1808 season started a run of lucrative whaling years for the British. In 1812 the value of the British whaling fleet was £1,500,000 and its 138 ships were employing about 6000 men. However, already involved in the Napoleonic Wars, Britain came into conflict with the United States from 1812 to 1815. This war inconvenienced the Arctic whalers only through the possibility of attack from the American privateers *Rattlesnake* and *Scourge*, but it was a different story for the British whalers sailing to the Southern Whale Fishery. Captain David Porter, commanding the United States frigate *Essex*, dealt a severe blow to this branch of the British whaling industry when he destroyed twelve British whalers in the Pacific.

Peterhead, Scotland in 1813. Peterhead was one of Scotland's leading whaling ports which kept the British whaling trade going after the English ports had ceased to follow the trade.

Author's collection

6 American Expansion: British Decline

Although the American whaling industry was in a precarious position when peace was declared in 1815 the prospects were bright and the whaling merchants were quick to seize the opportunity. In 1815–16 Nantucket despatched 48 whaleships and New Bedford 19 and it was largely owing to the example set by these two ports that American whaling survived and prospered so rapidly. Their success encouraged other towns to re-engage in or to enter the whaling industry as merchants realised they could develop a vigorous and profitable trade within the economic framework of an expanding, independent nation. Nantucket whalemen living in self-enforced exile still maintained a strong tie with their island home, and as the prospects in Dunkirk and Milford Haven diminished many of them returned. Others gave up their service with British whaling companies to take up the trade from the island in a United States where increasing prosperity was bringing a bigger demand for whale products.

Whilst some vessels continued to visit the Arctic it was the Southern Whale Fishery which attracted nearly all the growing fleet. At first the whalers which ventured round Cape Horn into the Pacific concentrated on taking whales close to the shores of South America, but in 1818 the *Globe* of Nantucket sailed into the open Pacific. Her captain, George W Gardener, reported sighting numerous whales and this enticed many whaling ships into the area 5°–10°S, 105°–125°W. This became known as the 'Offshore Ground' and within two years fifty vessels were there.

Following a report from Captain Winship of Brighton, Massachuesetts, that he had seen sperm whales in huge numbers off the coast of Japan, a Nantucket ship, the *Maro*, commanded by Captain Joseph Allen, and an English vessel, the *Syren*, captained by a Nantucketer, Frederick Coffin, sailed to Japanese waters in 1820. The ships returned filled with oil and two years later between thirty and forty vessels were taking whales in that area. Whaling grounds stretching from the Persian Gulf to Madagascar were opened up in 1823 by the *Swan*, owned by the English firm of Enderby. Most of North America's

The South Sea whale-fishery. The whalers *Amelia Wilson* and *Castor* hunting the sperm whale off Bouro, painted by W J Huggins, Marine Painter to His Majesty King George IV.

Author's collection

West Coast was traversed by the whalers fron East Coast ports and, in 1828, four Nantucket ships led the way to the whaling grounds off the coast of Zanzibar and into the Red Sea. In 1833, 392 American vessels, crewed by 10,000 men, sailed to hunt the whale. These ships and their equipment were estimated to be worth $12,000,000 and were bringing in an annual income of around $4,500,000.

The whaling industry had become an important contributor to New England's economy. Many people owed their livelihood to it and many trades profited by it. Apart from employment directly concerned with whale products, vessels had to be built and fitted out, equipment manufactured, maintenance and repairs undertaken, and whaling crews kitted out and given board and lodging in port. Other services, some dubious, were provided in whaling ports for men on shore leave. It has been estimated that, including owners, agents, insurance brokers and crews, 70,000 people were involved in some way with the whaling industry, a number which increased as the peak period was reached.

Apart from the early days when Indians, and later some negroes, supplemented the crews, whaling ships were traditionally manned by white New Englanders, who made whaling their livelihood and eventually, with the acquisition of skills and experience, a profession in which the family and the whole community took pride. In the late 1820s, however, the supply of these whalemen could no longer meet the demand of larger vessels, and the owners had to look elsewhere for crews. Men of all nationalities, races, creeds and character were drawn to the whaling ports. They came for all sorts of reasons: money, adventure, to escape the law; there were those who got drunk and woke too late to the sound of the swishing sea and the creak of timbers; there were black sheep of well-to-do families; there were beachcombers and drifters.

Above 'The *Brunswick*'. This ship of 357 tons first sailed on a whaling voyage from Hull in 1814.

By courtesy of Kingston-upon-Hull Museums

Below The *Esk*, commanded by William Scoresby Junior, sailed from Whitby on 29 March 1816. She was badly damaged in the ice at the end of June, and it was thought that she might have to be abandoned. However, with help, especially from the crew of the *John* of Greenock, commanded by Thomas Jackson (Scoresby's brother-in-law), the *Esk* was saved and reached her home port on 27 July. The picture shows an attempt to careen the ship, to bring the keel to the surface so that the damage could be repaired. This attempt was unsuccessful but temporary repairs saved the ship. When the *Esk* and *Lively*, also of Whitby, were lost in 1826, a total of 65 men perished.

From 'An Account of the Arctic Regions' by W Scoresby

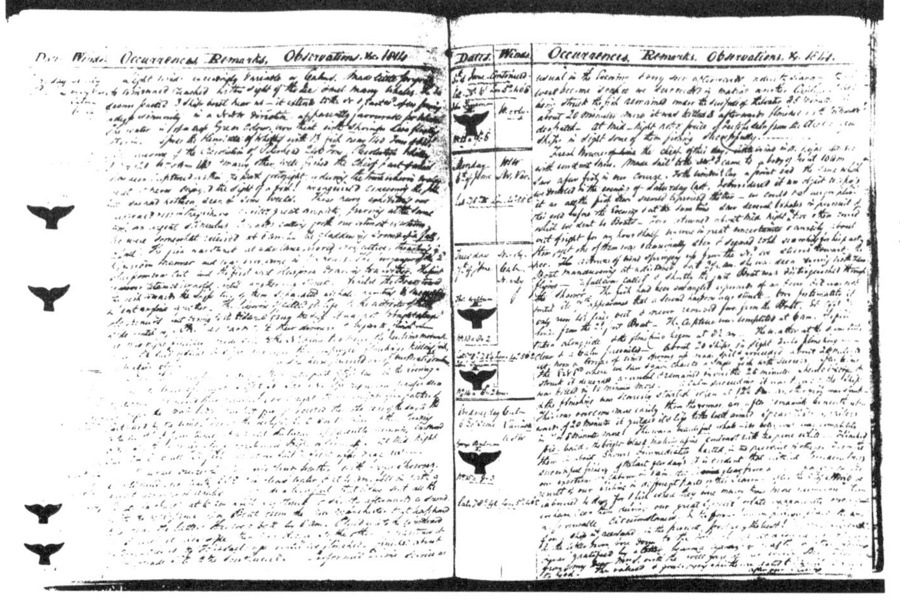
Date. Winds. Occurrences, Remarks, Observations, &c. 1814

Dates. Winds. Occurrences, Remarks, Observations, &c. 1814.

The logbook of the *Esk* (354 tons). Built by Fishburn and Broderick of Whitby in 1812, this vessel sailed to the Arctic every year from 1813 to 1826, but returning to Whitby in 1826 she was wrecked at Marske on the North Yorkshire coast with the loss of 22 men.

By courtesy of the Whitby Literary and Philosophical Society

A drawing of the Greenland whale by William Scoresby Junior.

By courtesy of the Whitby Literary and Philosophical Society

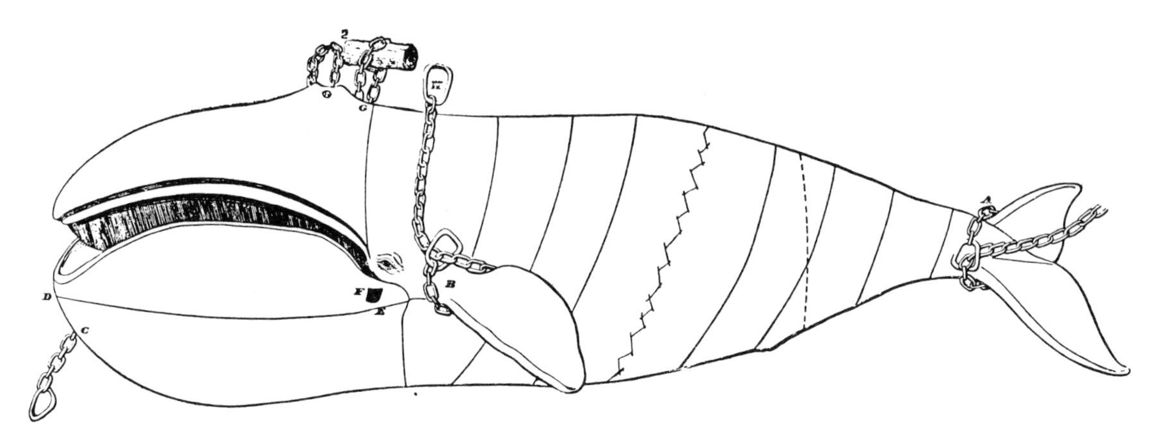

The method of cutting-in a right whale. The fluke-chain was attached at A and the fin-chain at B. The cutting tackle was hooked to the ring at H, and the whale could be hauled on to its side. A hole was cut at the root of the lip F and a cut made E-D. A blubber-hook was fixed in the hole and as the tackle was hauled tight the lip was severed from the jaw-bone and taken on deck. A cut was made just forward of the eye across the head and below the fin. A head-strap was fixed at GG using the blow holes. When the tackle was attached it was possible to sever the head from the body and haul it on board so that the baleen could be obtained. A throat-chain was attached at C and the throat taken on board as it was severed from the body. The first blanket piece was taken off near the fin and the rest of the body blubber was peeled off as far as the dotted line in the diagram. The whole flesh was cut through here, the stripped carcase allowed to clear the ship, the small stripped and the tail taken on board.

From 'The Marine Mammals of the Northwestern Coast of North America' by Charles M Scammon

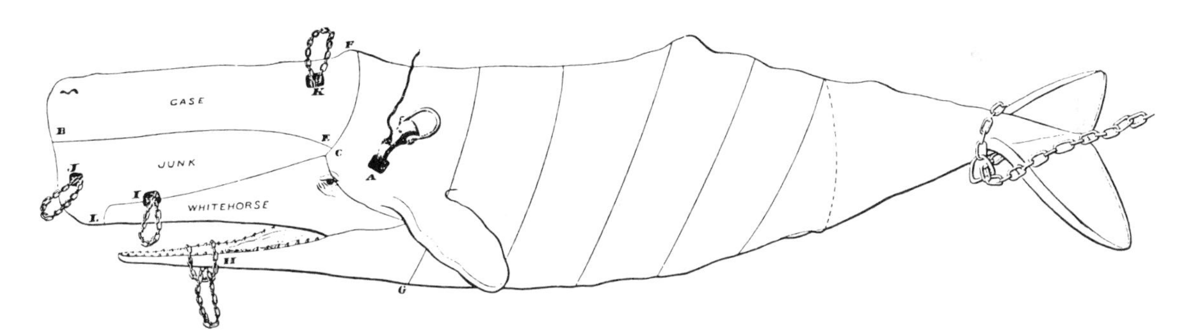

The method of cutting-in a sperm whale. The whale was fastened to the ship by a fluke-chain and head-rope. A hole was cut between the eye and fin at A. Cuts were made at each side of this section so that after a blubber-hook was inserted at A this section of the blubber could be raised by means of tackle and the whale's body rolled over. Cuts were made L-C between the upper jaw and the junk, B-E between the junk and case, E-F at the root of the case, corner of mouth to G. A chain-strap was fixed to the lower jaw H and attached to the second cutting tackle. This was raised and by manoeuvring the two tackles the whale was rolled over on its back. The lower jaw was severed from the head and taken on deck. The tackle attached to A now turned the whale so that similar cuts could be made on the opposite side. Holes were made at I, J and K. Straps were inserted and lines made fast. The second cutting tackle was attached to I, the fluke-chain slackened off so that the carcase was raised to an almost vertical position. C-L and E-F were completely cut away so that the junk and case were free from the body. With the head fastened to the ship the fluke-chain was hauled in and the carcase secured so that the flensers could strip the blubber. They cut spirally round the body as the tackle fastened to A was hauled. Thus the blubber was unwound from the body until the small was reached. The tail was cut off and the rear end of the body taken on deck. The head was taken on deck by using the tackle at J, the junk and case were separated at B-E and the spermaceti baled from the case. If the whale was a large one the junk and case were separated as it hung alongside the ship. The junk was taken on board, the case raised to the level of the deck and baled there.

From 'The Marine Mammals of the Northwestern Coast of North America' by Charles M Scammon

Inevitably these newcomers did not have the same interest in whaling, which was not always to their liking after their first taste. This resulted in desertions, and ships on a four-year voyage rarely had the same crews as they set out with when they returned to their home ports. On their way to the Pacific, American whalers often called at the Azores and Cape Verde Islands to pick up Portuguese, who were good boatmen and keen lookouts. South Sea Islanders, who were quiet and capable and made good crewmen, were also taken on. They had to get a special licence from the governor of the island on which they lived before signing on, while the captain of the vessel was under an agreement to return them to their home island within three years or pay a fine.

The owners of whaling vessels often employed agents to recruit their ship's crews. Unscrupulous agents would sign on any man in order to fulfil their commissions and get paid; they might also extract money from the men they took on. Most whaling captains were above reproach, but inevitably there were some of less admirable character who would even sell goods from the ship's chest to the crew at a high price. The items might have to be taken on credit, with the amount deducted from the men's wages at the end of the voyage. With a poor catch or a drop in the price of oil, the share from the proceeds would be low and it was not unknown for a man to be in debt or have very little money on his return from a voyage, in which case all he could do was to sign on again hoping for better luck. For those more fortunate a fat wallet awaited them when they went ashore. There were many temptations to 'let go' after a long absence at sea, and land-sharks and prostitutes were eager to relieve them of their money. Such men, finding themselves broke within a short time of being in port, could do little else but sign on for another voyage, even if it was on a different ship.

Long hours of boredom on the way to the whaling grounds, a bad captain, bullying officers, atrocious food and conditions and a poor catch could all create unrest among a discontented crew and might lead to desertions and even mutiny. Captain Neale of the *Reaper* out of Salem found himself with a crew which gave him enough trouble to cover six voyages. He was obliged to concede to their wishes to put into port, where he anchored out in the harbour and permitted drink and girls to be brought on board, fearing wholesale desertion if he allowed the crew to go ashore.

The decorated taffrail piece of an American whaler.

By courtesy of the Seaman's Bank for Savings in the City of New York

'Peche de la baleine', a French whaling scene by Chardon of about 1825.

National Maritime Museum, Greenwich

The American whaling trade was not completely given over to such riff-raff. There were many honest, godfearing seamen among the crews, interested in their work and bringing to it a courage and skill to match any in the history of the sea. There were considerate captains, good officers and happy crews, men who took good care of their wives and families and returned to whaling, voyage after voyage, because it was in their blood.

The captain of a whaling vessel was in complete command when at sea, and back in port he was answerable only to the owner and the courts when matters relating to the ship were in question. He had to be a skilled navigator and he had to know about whales and where to find them. He had to be an expert in the technique of catching, killing and flensing the whale. He might take charge of one of the whaleboats during the pursuit of the prey, but if he chose to remain on board ship he directed the general strategy of the hunt. He had to be able to handle men and exert his authority. As American whalers did not carry a surgeon, his duties included that of doctor, and more often than not his medical knowledge came from a book.

The mates, answerable to the captain, supervised all the work on board and were therefore responsible for the efficiency of the ship. Discipline enforced through the mates reflected the character of the captain, although a mate could put his own stamp upon it. The type of discipline was often determined by the type of crew, many of whom answered only to a strong, heavy-handed regime. The mates had also to be experienced whalemen, for, apart from steering the whaleboats, it was their job to attack the whale with the lance after it had been harpooned. Once the dead whale was alongside the ship, they took charge of the flensing, cutting and boiling.

The next most important men in the crew were the harpooners, upon whose skill and experience so much depended. The cooper was responsible for the construction of the casks and for keeping them in good repair to prevent any loss of oil. If the captain was in one of the whaleboats taking part in the chase and there was no shipkeeper on board, then the cooper took charge of the ship and directed it during the hunt. Other specific tasks were carried out by the carpenter, blacksmith and cook, while the steward, apart from being the personal servant of the captain, also waited at the officers' table and was in charge of the officers' and harpooners' provisions.

The remainder of the crew were foremast hands, among whom were experienced seamen and green hands. They carried out all the routine work of running a whaleship and dealing with a dead whale. Four of these men were allocated to each boat as rowers. The foremast hands were quartered in the cramped space of the fo'c's'le, with the harpooners, the steward and sometimes the cooper or shipkeeper amidships in what was known as steerage, while officers were housed aft in cabins.

In spite of its hazards and hardships, whaling life offered thrills and excitement to men with wanderlust and spirit of adventure. Aboard the whaleships, the green hands could learn seamanship in the best school in the world. Youngsters keen to make a successful career at sea joined the whalers to gain experience and promotion and were off to a good start by the time they transferred to a merchantman.

Attacking a sperm whale.

Author's collection

'The *Swan* and *Isabella* in the Arctic regions'. Painted by John Ward.

By courtesy of Kingston-upon-Hull Museums

A CORRECT STATEMENT

OF THE

Success of the Hull Ships at the Greenland and Davis' Straits Fisheries,

In the Year 1821.

GREENLAND.

Ships' Names.	Register Tonnage. tons	pts.	No. of Men.	Captains' Names.	Date of arrival at Hull	No. of Fish.	Actual weight of Fins or Whale Bone. tons.	cwt.	qr.	lb.	Actual quantity of Oil boiled. tuns.	qr.	gall.
North Briton	262	,,	41	John Allen	Aug. 2	10	6	16	0	7	167	3	29
Perseverance	251	,,	47	Matthew Wilburne	— 10	10	8	6	3	13	151	1	9
Everthorpe	349	,,	47	Robert Ash	— 16	11	9	19	1	12	182	1	30
Cicero	325	5	41	William Leaf	— 21	7	4	13	2	1	101	1	47
Mercury	346	25	48	William Jackson	— 21	13	7	14	3	20	161	3	41
Elizabeth	321	,,	45	Thomas Rhoades	— 22	10	5	16	2	20	135	0	36
Truelove	293	70	43	Thomas Todd	— 23	3	2	17	1	7	54	2	6
Walker	335	,,	49	Richard Harrison	— 25	7	5	14	0	26	105	0	24
Laurel	321	,,	41	Edward Dannatt	— 25	7	4	5	1	7	96	0	34
Manchester	285	,,	46	John Mitchinson	— 25	5	5	16	1	0	39	2	59
Shannon	348	,,	41	Robert Kelah	— 25	4	2	11	3	26	5[illegible]	3	33
Ebor	283	23	49	Thomas Lee	— 26	5	4	8	0	11	8[illegible]	3	35
Venerable	328	,,	41	John Bennet	— 27	10	4	17	0	17	[illegible]	0	4
William Torr	280	80	46	Phillip Dannat	— 27	5	3	2	0	16	6[illegible]	2	57
Dordon	285	90	49	William Gilyott	— 27	4	2	14	1	5	5[illegible]	0	22
Neptune	356	17	49	Martin Munroe	— 30	5	4	2	1	[illegible]	7[illegible]	[illegible]	[illegible]
Alfred	303	,,	49	William Clark	— 30		*Clean.*						
Gardiner and Joseph	360	,,	41	James Angus	— 31	7	2	19	3	10	8[illegible]	1	38
Exmouth	321	21	41	Edward Thompson	— 31	1		2	2	18	[illegible]	0	24
Cyrus	346	48	42	William Beadling	— 31	7	4	9	1	22	9[illegible]	0	11
Mary and Elizabeth	317	27	34	Robert Williams	Sept. 1	1		6	2	11	[illegible]	2	51
Abram	319	14	48	William Harrison	— 1	3	2	13	0	8	5[illegible]	0	55
Trafalgar	330	,,	46	William Lloyd	— 1	7	4	3	1	3	9[illegible]	3	41
Duncombe	270	,,	45	John Corbett	— 13	9	6	3	3	2	13[illegible]	3	4
Eagle	289	,,	49	William Brewis	— 13	14	7	13	1	3	15[illegible]	0	58
Harmony	300	,,	45	Charles Sawyer	— 15	3	2	10	2	27	5[illegible]	2	24
Fame	377	12	56	William Scoresby	— 19	9	6	12	3	22	14[illegible]	3	15
Jane	359	,,	41	Stephen Gamblin	— 20	1		18	3	9	17	2	37
Cato	305	,,	41	Andrew Turnbull	— 20	1	*No Bone.*				5	1	46
Unity	272	81	35	William Short	— 22	11	5	18	3	21	117	2	54
Rachel and Ann	223	84	40	Richard Marshall	— 25	14	7	0	3	23	14[illegible]	3	28
Total. 31 Ships	9665	33	1376			204	135	10	3	15	2745	2	7
Average each Ship	311	73 90/31	44 12/31			6 18/31	4	7	1	22 3/31	88	2	16 15/31

DAVIS' STRAITS.

Ships' Names.	Register Tonnage. tons	pts.	No. of Men.	Captains' Names.	Date of arrival at Hull.	No. of Fish.	Actual weight of Fins or Whale Bone tons.	cwt.	qr.	lb.	Actual quantity of Oil boiled. tuns.	qr.	gall.
Ellison	357	61	47	John Johnson	Sept. 19	13	10	15	0	15	170	3	56
Cumbrian	375	34	55	John Johnson	Oct. 4	28	12	11	1	25	213	3	34
Zephyr	342	,,	41	John Unthank	— 6	19	14	0	0	0	213	2	52
Gilder	360	7	41	George Bruce	— 6	19	12	19	0	23	219	0	33
Brunswick	357	,,	52	William Blyth	— 7	21	11	15	3	12	269	0	30
Andrew Marvel	377	,,	41	Thomas Orton	— 13	18	13	1	1	15	215	2	61
Albion	321	61	48	Richard Humphreys	— 17	21	12	15	1	7	214	1	39
Kirk Ella	410	72	49	Henry Watson	— 17	6	5	1	3	13	80	1	57
William	350	,,	46	Thomas Hawkins	— 17	12	7	15	3	10	129	1	47
Lee	363	,,	41	Thomas Forster	— 17	11	8	14	2	20	123	2	4
Kiero	358	,,	46	James Colquhoun	— 17	4	3	10	3	20	53	1	24
Egginton	336	27	47	John Wilson	— 17	8	6	17	2	4	114	2	17
Lord Wellington	354	,,	41	John Boydon	— 18	15	10	4	3	26	144	1	52
Progress	307	56	49	Matthew Mercer	— 18	8	5	5	3	16	97	2	57
Mary Frances	385	73	41	Thomas Wilkinson	— 18	14	8	4	2	6	161	0	21
Jogria	316	,,	49	James Mackintosh	— 19	[illegible]	[illegible]	8	1	[illegible]	[illegible]	[illegible]	[illegible]
Royal George	366	28	49	Joseph Peckit	— 20	9	5	9	1	19	108	0	0
Friendship	410	,,	49	George Green	— 21	6	4	1	2	21	80	0	15
Ariel	340	,,	49	William Hurst	— 21	12	6	7	0	15	118	0	23
Thomas	355	,,	41	William Brass	Nov. 6	10	7	10	1	7	128	3	1
Margaret	339	12	41	James Creighton	— 6	8	4	1	3	4	76	2	46
Total. 21 Ships	7482	55	963			294	186	19	1	12	3142	3	6
Average each Ship	356	29 10/21	45 18/21			14	8	18	0	7 5/21	149	2	39 0/21

GRAND TOTAL AND AVERAGE FOR GREENLAND AND DAVIS' STRAITS.

	Register Tonnage. tons	pts.	No. of Men.	No. of Fish.	Bone tons.	cwt.	qr.	lb.	Oil tuns.	qr.	gall.
Total. 52 Ships	17147	88	2339	498	322	10	0	27	5888	1	13
Average each Ship	329	72 80/52	44 51/52	9 30/52	6	4	0	4 1/52	113	0	59 37/52

☞ The Refuse, or Black Oil, is not included in this Account, nor the number of Men taken from Shetland and the Orkneys.—The John, 343 Tons; Symmetry, 342 Tons; Harmony, 378 Tons; Leviathan, 409 Tons; Henry, 314 Tons; Cervantes, 356 Tons; Aurora, 368 Tons, Lost at Davis' Straits, Crews saved.—Thornton, 262 Tons, Lost at Greenland.—Hebe, 364 Tons, Lost on her passage to Davis' Straits, Crews saved.

AGGREGATE STATEMENT OF THE NUMBER OF THE HULL SHIPS FROM THE GREENLAND AND DAVIS' STRAITS FISHERIES, FROM THE YEAR 1772, WITH THE QUANTITY OF OIL, &c.

Year.	No. of Ships.	Tuns of Oil.		Year.	No. of Ships.	Tuns of Oil.		Year.	No. of Ships.	Tuns of Oil.		Year.	No. of Ships.	Tuns of Oil.		Year.	No. of Ships.	Tuns of Oil.	
1772	9	391		1782	3	217		1792	20	896	3 Clean	1802	34	2955		1812	49	6812	1 Captured
3	9	265	2 Clean	3	4	290		3	18	835	1 Clean	3	41	2262	2 Clean	3	54	3533	3 Clean
4	8	466	1 Lost	4	9	432		4	16	709	1 Lost	4	40	4018	3 Lost	4	57	7379	1 Lost
5	10	68	6 Clean and 2 Lost	5	15	722	1 Clean	5	14	1148	1 Clean	5	38	5165	1 Lost and 1 Captured	5	55	3607	3 Clean, 1 Lost
6	9	275	1 Lost	6	19	856	1 Clean and 1 Lost	6	17	1578	1 Captured	6	39	3524	3 Captured	6	55	5150	1 Clean
7	9	333		7	29	1132	1 Lost	7	21	1741	1 Lost	7	35	4350	2 Lost	7	57	4789	1 Clean, 1 Lost, 1 Broken up
8	8	171	2 Clean	8	34	958	3 Clean	8	23	2162		8	27	4556	3 Lost and 2 Captured	8	63	6219	1 Lost
9	3	142	1 Lost	9	27	854	2 Clean and 2 Lost	9	26	2244	2 Lost	9	26	4321	3 Lost	9	60	5077	3 Lost, 2 Broken up
1780	4	309		1790	23	832	2 Clean and 1 Lost	1800	22	1818		1810	34	5019		1820	60	7978	2 Lost, 1 Broken up
1	3	261		1	18	345	4 Clean and 3 Lost	1	24	2149	1 Lost	1	42	5398	1 Lost,	1	52	5888	1 Clean, 9 Lost

☞ The lost and captured Ships are not included in this number of Ships.

A statement of the success of Hull ships in the Northern whale-fishery, 1821.

By courtesy of Kingston-upon-Hull Museums

The whalemen of every nation were explorers as well as hunters, and the discovery of numerous islands in the Pacific can be attributed to them. In many cases they were the first strangers to make contact with the natives. Unfortunately the best relationships were not always established and some of the blame for the exploitation of the natives can be placed on whalemen as well as on merchant sailors. This led to retaliations, and there are many records of antagonism between whalers and natives. Some of the most unfortunate incidents occurred in New Zealand, where, apart from the continual visits by the deep-sea whalers, shore stations were set up. The first of these was started at Preservation Inlet in 1827, although it had no real success until 1829. The number of shore stations increased to a point where overkilling occurred, so that the end of bay whaling was in sight by 1836.

However, the enterprising merchants of Hobart, Tasmania, had already started to build deep-sea whalers, the first of which, the *Caroline*, sailed in 1829. From then on whaling played a significant part in Hobart life. Alexander McGregor owned most whalers throughout Hobart's whaling era but the most successful were the Bayley Brothers. One of the most colourful characters to enter the Tasmanian whaling trade was Captain James Kelly. In 1821 he sent his brig *Venus* to hunt whales on the edge of the pack ice. The brig went as far as 72°S but found the hazards too great and Kelly did not repeat the experiment.

The British levied a duty of £26 12*s* per ton on all whale oil from foreign countries but only 1*s* on that taken under a British flag. This helped Australian and Tasmanian whaling, and practically all the oil obtained by Tasmanian whalers was sent to Britain. In the 1840s the Tasmanian whaling fleet consisted of 40 vessels, with 200 boats and about 1000 men employed on them. Bay whaling had also extended along the coast of Australia, and in many cases British and American whalers combined deep-sea and offshore whaling from stations which were established along the south and south-west coasts.

Following the end of hostilities in 1815, the British virtually had the northern whaling seas to themselves. A few American vessels went to the Arctic but the Dutch and German ships were devoting most of their time to sealing, and apart from the French, who mounted some small expeditions, other Europeans had little interest. The outlook for the industry was promising.

In 1817 the *Larkins* of Leith and the *Elizabeth* of Aberdeen carried out some exploration and reported many whales on the west side of Baffin Bay. The following year the Admiralty organised an expedition to search for the North-west Passage and made use of the knowledge gained by these two whalers. This was reciprocated when the naval expedition showed that the whalers could work their way south along the west coast of Baffin Bay and Davis Strait with a very good chance of finding a plentiful supply of whales in the numerous bays and inlets which indented the coastline and formed a maze of waterways.

The harpoon-gun of the *Truelove*.

By the courtesy of the Wardens and Brethren of the Corporation of the Hull Trinity House

Manifest of the Cargo of the Ship Baffin, British built, burthen three hundred twenty one 87/94 tons, William Scoresby Junr Master from Greenland, of and for Liverpool:- To wit,

Two hundred forty leagers.
Sixty puncheons.
Three Barrels.
} in all, three hundred three Casks, containing five hundred and fifteen butts of blubber of half a ton each.

Eight tons of whale-fins, not packed.
One seal skin.
Five bear skins.
Three bears.

} To Messrs N. Hurry & Gibson.

The above being the Produce of Animals taken in the Greenland Seas,
Consisting of Seventeen Whales, one Seal, eight Bears.

Greenland Seas, 31st July, 1820.

William Scoresby Junr

The manifest of the *Baffin* for the voyage of 1820.

By courtesy of the Whitby Literary and Philosophical Society

In 1819 a record number of 65 ships left Hull, but although plenty of whales were seen the weather was so bad that few boats were able to chase with any success. The ice in Davis Strait was an even greater hazard and 10 British ships were lost there. Because of these disasters many captains returned to the Greenland Sea the following season when the Hull fleet of 60 ships brought back a record cargo worth £318,880. Many young whales were taken and this thoughtless killing meant that the stocks had no chance to recuperate and the fleet was forced to hunt in the more dangerous waters to the west of Greenland.

Throughout the years efforts had been made to develop a harpoon gun and once again, in 1820, experiments were carried out. They were no more successful than the earlier attempts and for nearly fifty years the hand-held harpoon remained the chief weapon for securing whales.

In the hazardous northern waters all sorts of circumstances played their part in determining results. The chief factor was the weather, which affected not only the condition of the sea and the ice but also the presence of whales. Skill and luck could mean a good return for a particular port or an individual ship, whereas others might meet with less success.

A model of the *Baffin*. The *Baffin* was designed by William Scoresby Junior and built at Liverpool in 1819. He took command of her the following year and made four successive voyages in her. She was lost in Davis Strait in 1830.

By courtesy of the Whitby Literary and Philosophical Society

The log of the *Baffin*.

By courtesy of the Whitby Literary and Philosophical Society

The practice by which vessels stayed out as long as possible in the hope of bringing back a cargo which would make the voyage profitable often meant a late return to port, and owners were finding that this jeopardised the chances of using the whalers for winter trading to the Baltic. Many were wondering if it would be more economical to abandon the whaling trade and concentrate on the Baltic timber trade. The situation differed from that in the American whaling industry where, once the Pacific was opened up, ships went on long voyages unhampered by seasonal limits and winter trading. British whaling was to continue for a considerable time but whalers which were lost were not being replaced and others were withdrawn from the service. Thus the number of ships leaving British ports to hunt the whale in the Arctic began gradually to decline: of the 92 ships which sailed in 1827 London sent only 2, and the 33 which left Hull was about half the number sent in that port's heyday. Scotland was sending more ships than England.

Then in 1830 tragedy struck the British fleet in Davis Strait. Of the 91 whalers sent out 19 were lost in the ice, nearly every vessel was damaged – some severely – and 21 ships returned without having made a kill. British whaling was at a low ebb and, in spite of good seasons in 1832 and 1833, the loss of 6 ships and poor returns in 1835 precipitated the end. Matters were not helped by the appearance of substitutes for whale products, and a series of poor seasons in the 1840s brought the end of the northern whaling industry nearer.

In 1849 the British whaling fleet had shrunk to 32 ships, with Hull, sending 12, the only English port still engaged in the trade.

The *Baffin* model, showing the flensing platform, from which men worked when flensing the whale. This platform, known as the cutting-stage, enabled the men to work without standing on the whale as they had done in earlier days.

By courtesy of the Whitby Literary and Philosophical Society

7 The Great Days of American Whaling

The *Charles W Morgan*, America's most famous whaler, started her first voyage on 6 September 1841 from New Bedford and returned from her final voyage to the same port on 28 May 1921; 20 of her 37 voyages were made from New Bedford and the other 17 from San Francisco. She is now preserved at Mystic Seaport, Connecticut, where she was taken in 1941.

By courtesy of the Seaman's Bank for Savings in the City of New York

The golden age of American whaling lasted for 25 years. It began in 1835, when the *Ganges*, out of Nantucket, captured the first right whale off the north-west coast of North America, on what was known as the Kodiak Ground – a most profitable addition to the American whaling industry.

The fleet had grown from 203 to 421 vessels in six years, and at this time nearly 30 ports were engaged in whaling. Some of them participated in other trades and ventures, but New Bedford built its reputation and wealth solely on the whaling trade. Certain ports became known for their specialisation within the trade. Nantucket concentrated chiefly on catching sperm whales; New London pursued the right whale, particularly in Davis Strait, while Sag Harbor and Stonington sent their ships to hunt the northern and southern right whale. Provincetown preferred sailing smaller vessels on shorter voyages into the Atlantic, but from New Bedford, the greatest whaling port of all, ships went anywhere if there was a chance of making a profit out of the whale.

The rapidly developing American nation provided a demand for whale products which the New England merchants were determined to meet. Everything was geared towards the expansion of a highly profitable industry. More and more sailing ships were built and the owners were anxious that the long voyage to the Pacific should prove worthwhile. In some cases they arranged for carrier vessels to meet their whalers, take the oil on board and return to New England, leaving the whaling ships to resume their hunt and fill their casks again. In other cases they looked to bigger ships to satisfy their desire for bigger cargoes.

In 1820 New England whalers were about 280 tons burthen but, with the extension of voyages, larger ships were called for, and by 1840 400-ton vessels were usual, having been found to be both suitable and an economic proposition. This move towards bigger whalers was partially responsible for Nantucket's decline. Because of a bar across the entrance, its harbour could be used only by smaller vessels of shallower

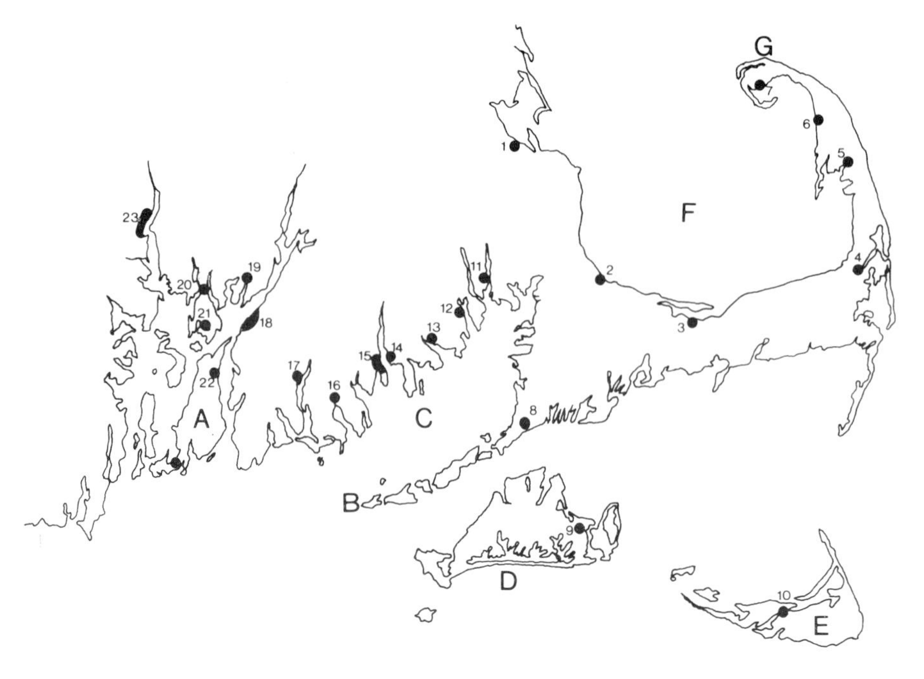

Some of the principal whaling ports of Massachusetts.
1. Plymouth
2. Sandwich
3. Barnstable
4. Orleans
5. Wellfleet
6. Truro
7. Provincetown
8. Falmouth
9. Edgartown
10. Nantucket
11. Wareham
12. Marion
13. Mattapoisett
14. Fairhaven
15. New Bedford
16. Dartmouth
17. Westport
18. Fall River
19. Swansea
20. Warren
21. Bristol
22. Portsmouth
23. Providence
24. Newport

Geographical Features
A. Rhode Island
B. Cuttyhunk
C. Buzzards Bay
D. Martha's Vineyard Island
E. Nantucket Island
F. Cape Cod Bay
G. Cape Cod

draught which could not compete economically with the larger ships. Attempts to make the harbour accessible, such as the use of a camel (a buoyant apparatus attached to a ship's sides to raise her in the water so that she could float over the bar), met with little success and the island's whaling trade began to run down steadily.

The whaling vessels were heavy and tub-like, often provoking cries of derision from other seamen. They were not built for speed and lacked the graceful lines of many merchant ships. But there was something attractive about a ship built, or altered, to serve in the whaling trade. She was constructed strongly for endurance, and her holds, designed to carry as many casks as possible, often gave the vessel a bulging appearance. She was square-rigged for deep-sea work. The barque became popular, as it could be easily handled by about six men when hove-to while the rest of the crew were taking to the boats to chase the whale.

Unless it was being used in the Arctic, the American whaler, unlike its British counterpart, had no need for reinforcement, although the bottom was covered with cedar ⅞in thick and encased in copper to protect it against rot and fouling by weed and worms. If the whaler was bound for the Arctic, then its bows were double-planked. The cutting spades used for stripping the whale played such havoc with the deck that it frequently had to be repaired by the ship's carpenter and was always relaid before the next voyage.

Peculiar to whaling vessels were the wooden davits, which were found more suitable for suspending, lowering and hoisting the whaleboats. Another distinctive feature of the American whaler, and also of British vessels which made the long voyage to the Pacific, was the try-works. This was a brick construction which held two huge iron try-pots above a fire. When trying-out was in progress, the heat was so great that a tank of water was placed below the grating holding the fire to prevent the deck from being scorched. No other vessel presented such a fiery spectacle, with flame and thick black smoke pouring from it; and when the work was carried out at night the whaler was an awesome sight.

The whaler was slow and ponderous with ungainly lines but her boats were just the opposite. Built specifically for the work in hand, the whaleboat was graceful and sleek of line, strong but light to ensure speed, and easily manoeuvrable. The size varied, but generally it was 28ft long and 6ft wide at the centre. It was sharp at both ends so that it could be rowed in both directions. American boats carried six men, one of whom was the boat-header, who steered with a 28ft oar. The others rowed. Each man sat the full width of the boat away from his rowlock; this was necessary so that he would be able to wield the oars, which were larger than any others used at sea and which would otherwise have been badly balanced. The American whaleboat narrowed towards each end from amidships, which meant that the rowers were seated at unequal distances from their rowlocks and used oars of various lengths: two short, two medium and one long. The lengths of these might vary from one boat to another but within each boat their lengths were such as to make an even balance: a typical complement would be oars 16ft, 17ft and 18ft in length. The long and two short oars were on the starboard side, with the two medium ones on the port side. Sometimes a sail

New Bedford in 1840. Built on the whaling trade, New Bedford became America's leading whaling port after the decline of Nantucket.

Author's collection

The *Ann Alexander* (211 tons) served in the merchant trade from 1805 to 1820 and then turned whaler, in which capacity she sailed from New Bedford until 1851 when she was rammed and sunk by a sperm whale in the Pacific. Two days later the crew were picked up by the *Nantucket*.

Author's collection

A model of the whaler *Alice Mandell*, 413 tons. Having left New Bedford on 10 August 1855 to hunt whales in the Pacific, she was lost on Prate Shoals in the China Sea in March 1857.

Crown Copyright, Science Museum, London

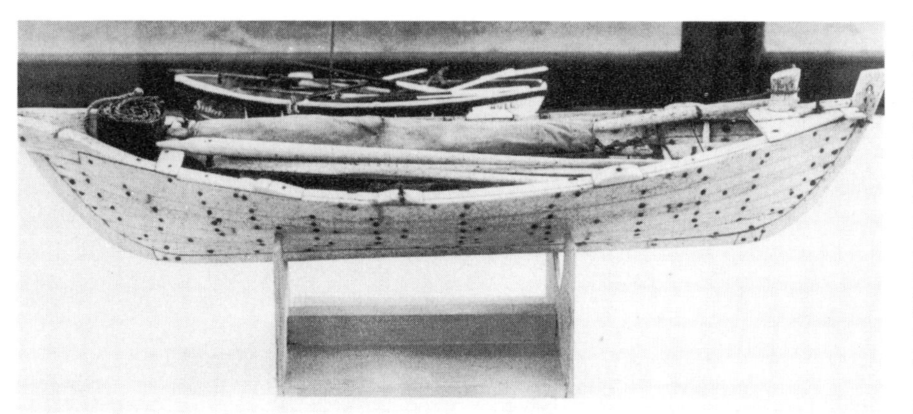

The British whaleboat was carvel-built and remained so because greater strength was required for work among the icefloes. In the early days they had square sterns, which disappeared in favour of the double-ender, especially after the capture of American whale ships during the American War of Independence. The standard crew was six, but five-, seven- and even eight-man boats were not unknown. The bollard (loggerhead) was in the bow whereas in the American boat it was in the stern.

By courtesy of Kingston-upon-Hull Museums

A 1/12 scale model of an American whaleboat. The early American whaleboat was clinker-built and remained so until the middle of the nineteenth century, when a composite clinker/carvel boat was introduced. The underwater section was carvel with flush strakes, with an inside covering stripped to each seam. The upper section was constructed on the clinker system of overlapping strakes. The boat was double-ended, and carried six men – five rowers and a helmsman.

Crown Copyright, Science Museum, London

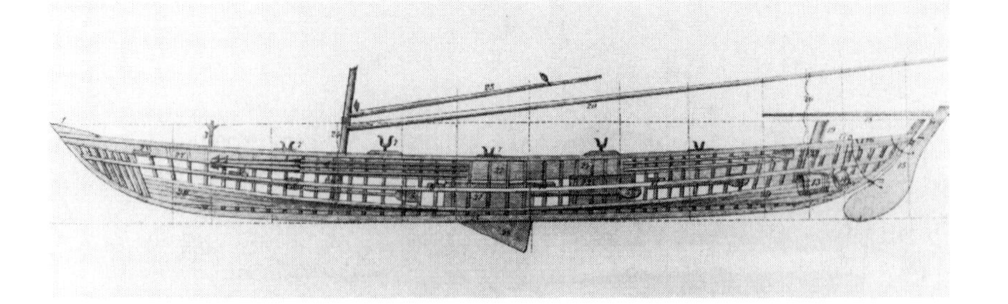

A sectional drawing of an American whaleboat.

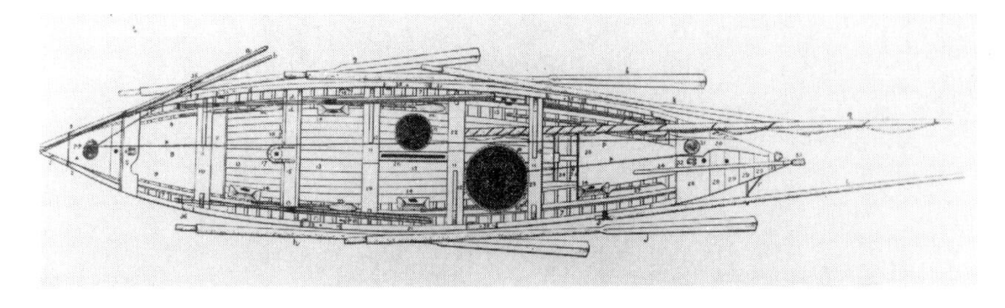

A deck view of an American whaleboat.

Implements belonging to a whaleboat.
1. Oar.
2. Boat-waif.
3. Boat-hook.
4. Paddle.
5. Boat-sails.
6. Sweeping-line buoy.
7. Lead to sweeping-line.
8. Chock-pin.
9. Short-warp.
10. Boat-piggin.
11. Boat-keg.
12. Lantern-key.
13. Sweeping-line.
14. Boat-hatchet.
15. Lance-warp.
16. Boat-grapnel.
17. Boat-knife.
18. Fog-horn.
19. Line-tub.
20. Boat-bucket.
21. Drag.
22. Nipper.
23. Boat-crotch.
24. Boat-compass.
25. Boat-anchor.
26. Row-lock.
27. Tub-oar crutch.
28. Hand-lance.
29. Pine-flued harpoon.
30. Toggle-harpoon.
31. Boat-spade.
32 & 33. Greener's gun-harpoon.
34. Greener's harpoon-gun.
35. Bomb-lance.
36. Bomb-lance gun.

From 'The Marine Mammals of the Northwestern Coast of North America' by Charles W Scammon

or paddles might be used instead of oars, as circumstances demanded, but, whatever the means of propulsion, the whaleboat was an almost perfect craft.

A vital piece of equipment was the whale-line, which had to take an enormous strain and so was made of the best tarred hemp. Not only the capture of the whale but also the lives of men depended on it, and the line was very carefully coiled and stowed in tubs so that it could be drawn out without fouling. The British stowed it in two tubs making for easier handling, whereas the Americans used one. Positioned near the bow of the whaleboat, the British line went round a bollard and over the bow; the American line ran to a loggerhead at the stern and then the full length of the boat. Both positions had their dangers and, through careless stowage or error of judgement, a harpooner could find himself whipped overboard by a snaking, coiling rope.

Some whaling was carried out in the Atlantic, but the majority of American vessels made for the Pacific, either by way of Cape Horn or round the Cape of Good Hope. Those taking the latter route left port in autumn and might call at St Helena and Tristan da Cunha before rounding the Cape. Then some sailed via the Seychelles and New Holland to New Zealand and Australia. Others sailed from the Cape direct to New Zealand, arriving not later than March. They spent six to eight weeks there and then followed one of two courses: cruising between 22° and 28°S to South America to begin the season on the Offshore Ground in November; or making for the Society Islands (to arrive in June) from where they went to Fiji and Samoa and back to New Zealand by March. The whalers which went to the Pacific via Cape Horn left their home ports in summer so that they could provision in Chile or Peru before November, when the season began on the Offshore Ground. After cruising in an area 5°N–10°S by 105°–125°W for up to three months, they

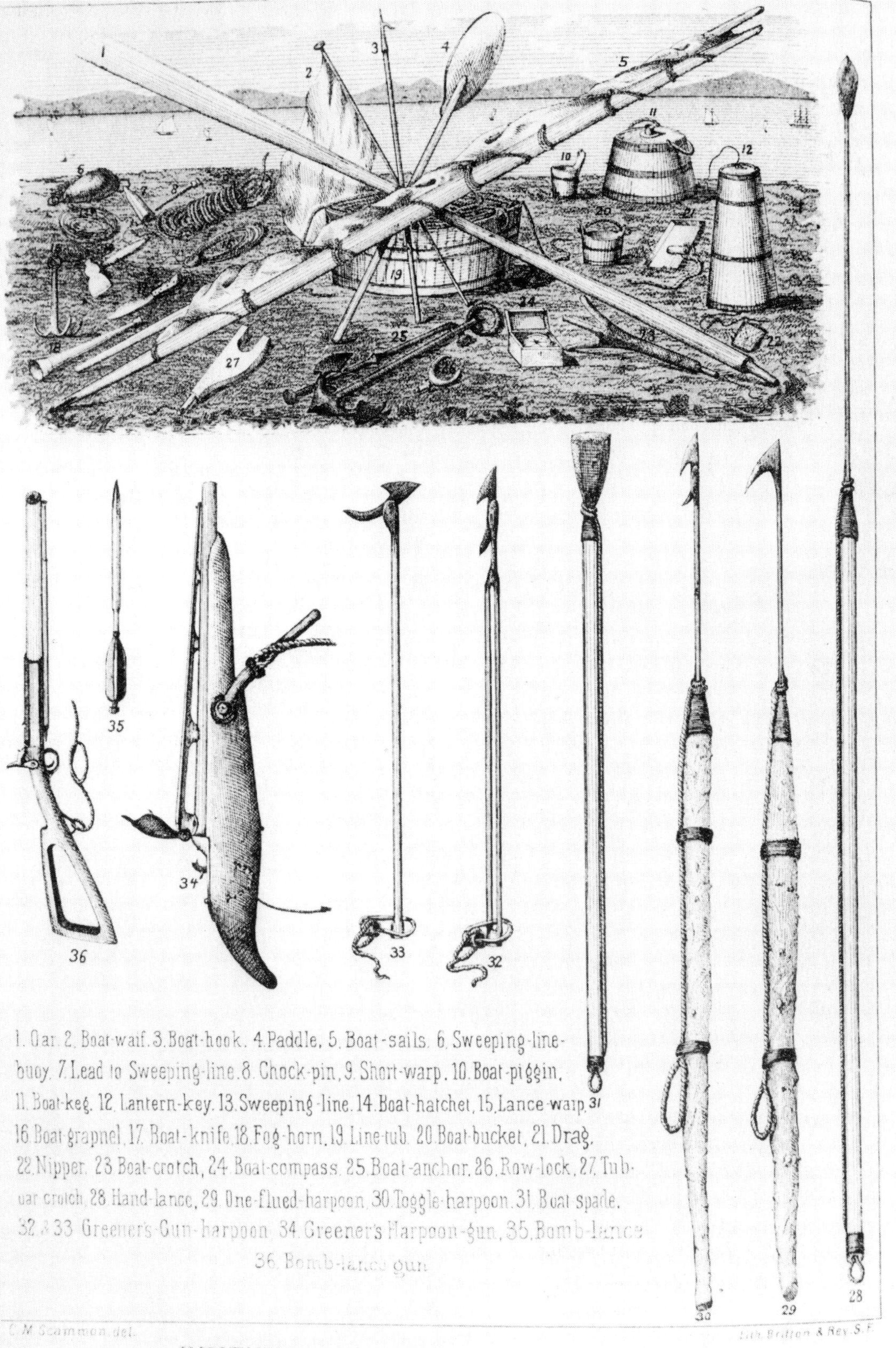

IMPLEMENTS BELONGING TO A WHALE BOAT.

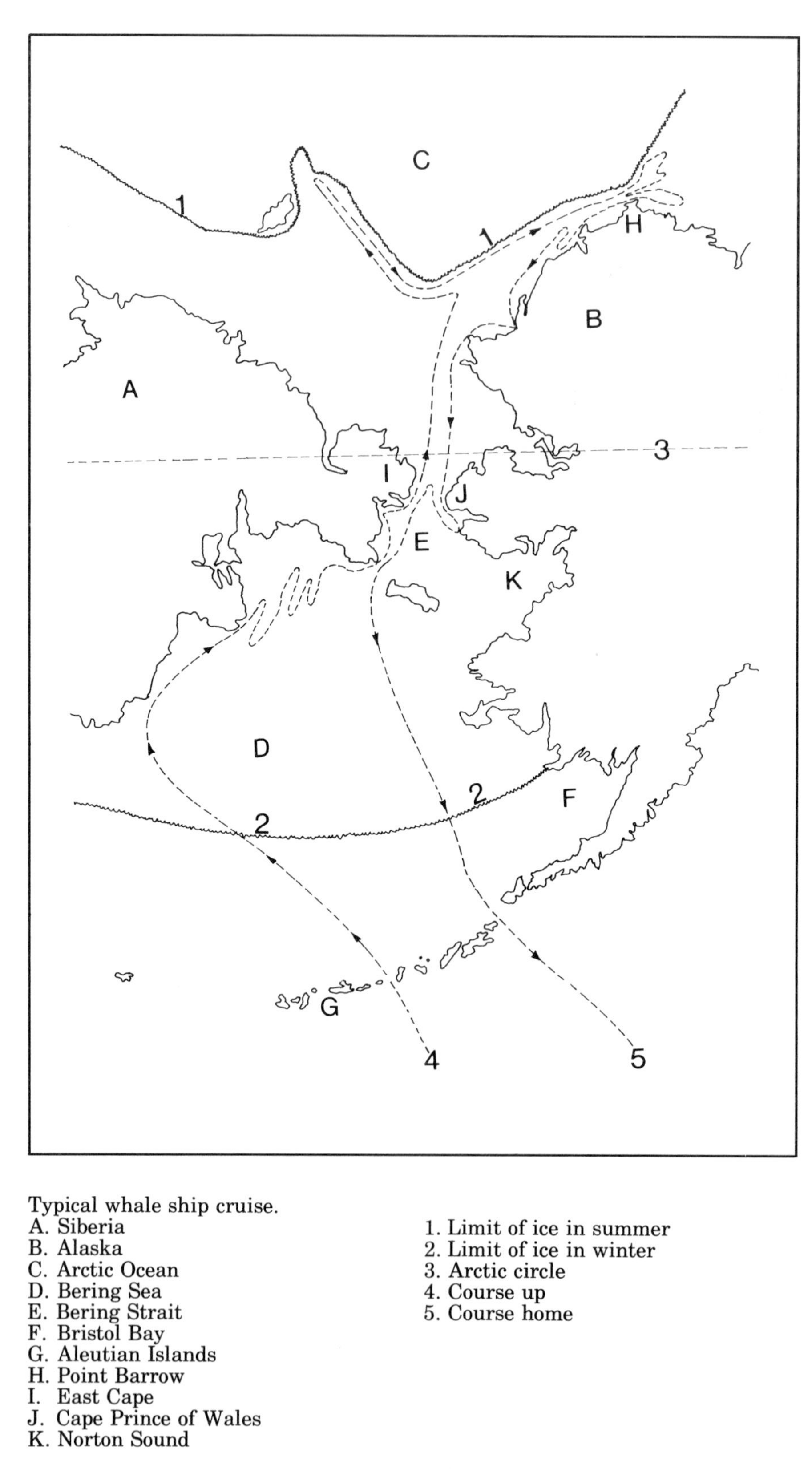

Typical whale ship cruise.

A. Siberia
B. Alaska
C. Arctic Ocean
D. Bering Sea
E. Bering Strait
F. Bristol Bay
G. Aleutian Islands
H. Point Barrow
I. East Cape
J. Cape Prince of Wales
K. Norton Sound

1. Limit of ice in summer
2. Limit of ice in winter
3. Arctic circle
4. Course up
5. Course home

moved north by the captain's chosen route. They sailed west along the equator to the Mulgrave Islands and finished off the coast of Japan, working their way towards the north-west coast of America. Alternatively, after leaving the Offshore Ground, they made for the Sandwich Islands, in the vicinity of which they spent the months of February, March and April, then sailing 30°N between 145° and 165°W. Whichever course they followed, the whalers timed their cruises so as to arrive in Honolulu in October, where they refitted for the next season on the Offshore Ground.

Once the Bering Strait had been successfully navigated by a whaler in 1848 a new pattern developed. Leaving their New England ports in the autumn, the whalers rounded Cape Horn to reach Honolulu in March or April. After provisioning, they sailed for the northern seas in the vicinity of and beyond the Strait. This 'regular season', as it came to be known, lasted only as long as the ice conditions permitted. With time to spare until the next regular season, the whalers carried out what they called their 'between-season whaling', cruising south to Honolulu and beyond into the South Pacific and returning to Honolulu in time to refit for the regular season.

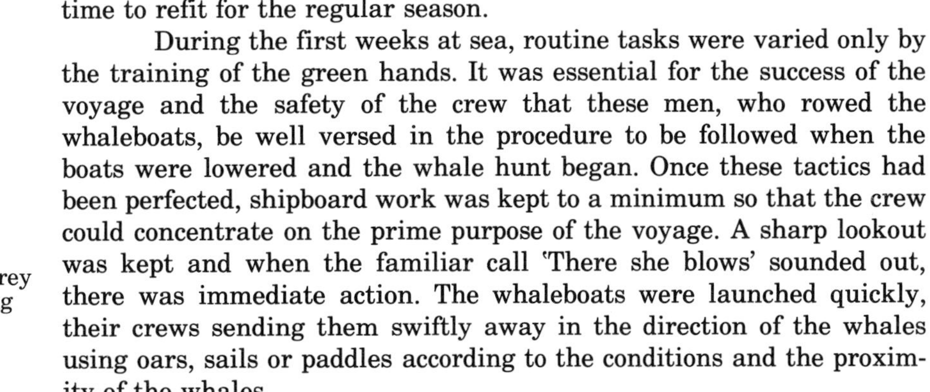

During the first weeks at sea, routine tasks were varied only by the training of the green hands. It was essential for the success of the voyage and the safety of the crew that these men, who rowed the whaleboats, be well versed in the procedure to be followed when the boats were lowered and the whale hunt began. Once these tactics had been perfected, shipboard work was kept to a minimum so that the crew could concentrate on the prime purpose of the voyage. A sharp lookout was kept and when the familiar call 'There she blows' sounded out, there was immediate action. The whaleboats were launched quickly, their crews sending them swiftly away in the direction of the whales using oars, sails or paddles according to the conditions and the proximity of the whales.

Attacking a right whale. In this aquatint by Ambroise Louis Garnerey (1783–1857) a whale is shown being flensed alongside the ship while another is being towed to it.

Author's collection

Attacking a sperm whale.

Author's collection

The whale strikes back.

Author's collection

Once the whaleboat had been manoeuvred into a suitable position, the harpooner forsook his oar and stood up ready to throw his harpoon. If the opportunity arose, he would follow this with a second harpoon, which was fastened by a short warp to the harpoon line. This was a precaution in case the first harpoon was pulled out. If this second harpoon was not thrown, it was discarded overboard, so that there was no risk of it fouling the line being run out by the whale.

Once the harpoon had struck, the order 'Stern all' was given, and the oarsmen reversed their rowing to take the boat as quickly as possible clear of the whale's immediate reactions. This was not an easy task and many a boat was overturned or smashed in these early moments. Such incidents were usually the result of the whale's violent convulsions after being struck and not of a deliberate attack on the boat such as might take place later in the 'fight'.

After being struck by the harpoon, the whale sometimes ran, towing the boat after it. This became known in the trade as the 'Nantucket sleigh-ride'.

Author's collection

When the whale was struck, it either 'ran' on the surface or it 'sounded'. If it ran, the whaleboat would be towed along with it – at up to 25mph in the early stages – and the men would be able to haul themselves nearer to the prey. If the whale sounded, the line was allowed to run out, and if the whale looked like taking all of it, a line from another boat was 'bent-on'. If the tub in the first boat was emptied of line before this could be done, then the line was severed to prevent the boat from being dragged under.

When the crew finally managed to shorten the line by hauling in, they brought the boat near enough to the whale for the boat-header, who had changed places with the harpooner, to use the lance. He struck the vital parts, and when he was satisfied that he had made a kill the boat was rowed to a safe distance, and the crew waited until the whale spouted blood and 'went into its flurry'. This was the term used to describe the furious swimming in an ever-narrowing circle. The end came with a violent thrashing of the sea before the whale rolled over on to its side, with its dorsal fin sticking up out of the water. The whalemen called this 'finning-out', a sign that the whale was dead. If the hunt was to be continued, the dead whale would be marked to show to whom it belonged, before being towed back to the ship.

Once alongside, it was secured to the starboard side of the ship with its head towards the stern, attached by ropes and a chain around the narrow part near the flukes. The cutting-stage, a narrow plank with a rail on the inside, was put in position over the side of the ship just beyond and above the whale. The cutters worked from the platform, using long-handled spades. The captain and first mate usually severed the head, whilst the second mate was responsible for the scarfing, which meant that he cut the blubber at an angle across the body, each cut being 12 to 18in from the next. Once the piece nearest the head had been eased up by the insertion of a hook and fixed to a block and tackle in the mainyard, it could be peeled from the body with the aid of the cutters. When the piece reached a certain size, depending on the height of the block, another hook was attached and the piece of blubber severed just above it so that the peeling could continue by means of the second hook, while the blanket piece, on the first hook, was lowered into the blubber room, situated between decks under the main hatch. Here the blanket pieces were cut into 'horse pieces', about a foot square. These were pitched up on to the deck for mincing. They were taken to a 'mincing horse', a small table secured to the rail of the ship where a boy with a short-handled hook held the piece steady while the mincer, with a two-handled knife, slashed it nearly through into thin slices, which just hung together. Because of their likeness to the pages of a book these pieces were known as 'Bible leaves'. They were put into try-pots and, after boiling out, the oil was passed into copper containers to cool before being transferred to casks. After being thoroughly cooled, these were stored in the hold.

Hazards of the whale-fishery.

Author's collection

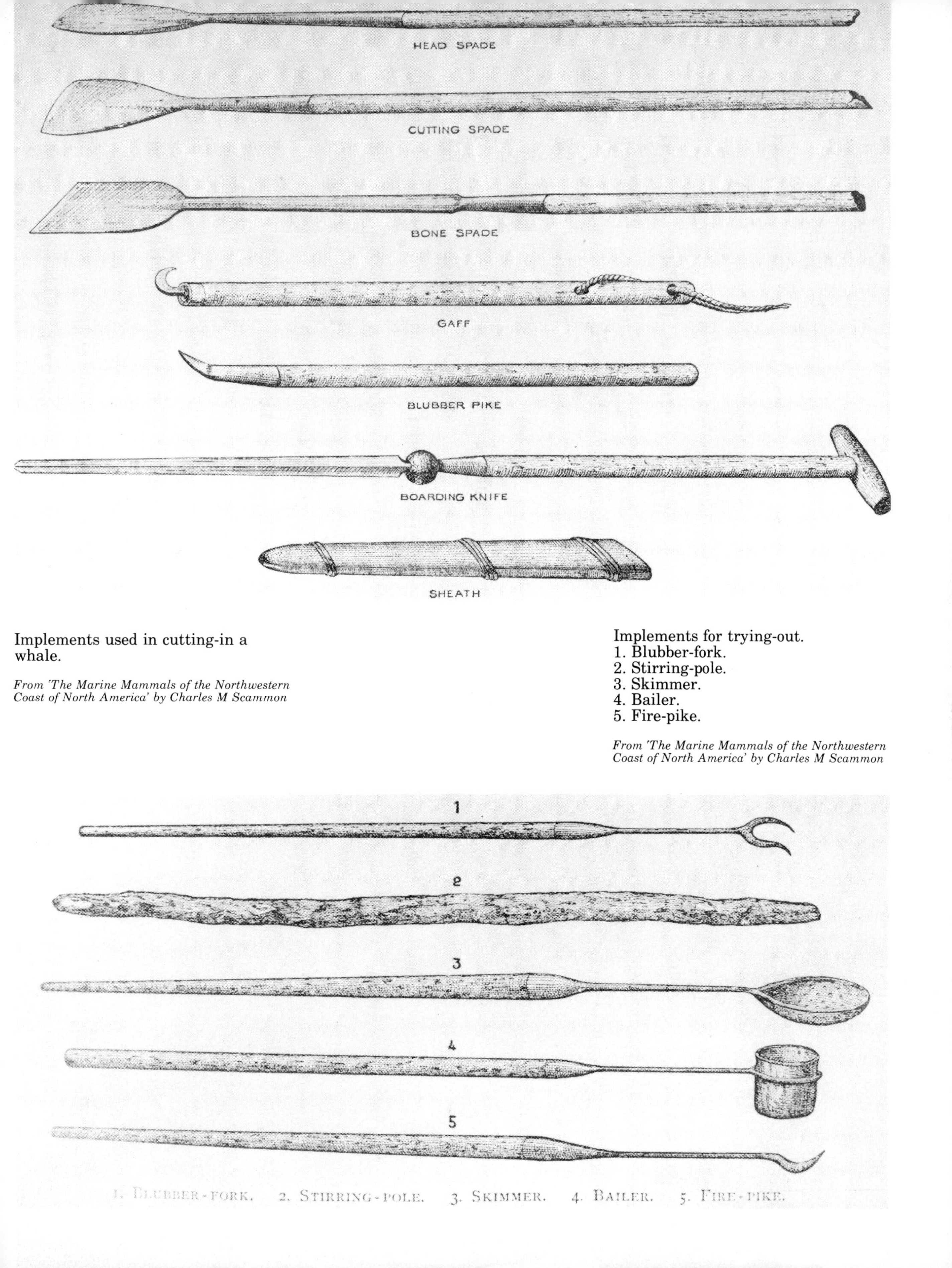

Implements used in cutting-in a whale.

From 'The Marine Mammals of the Northwestern Coast of North America' by Charles M Scammon

Implements for trying-out.
1. Blubber-fork.
2. Stirring-pole.
3. Skimmer.
4. Bailer.
5. Fire-pike.

From 'The Marine Mammals of the Northwestern Coast of North America' by Charles M Scammon

'South Sea whalers boiling blubber' by O W Brierly.

The Illustrated London News

This night scene is of the activity on board the famous American whaler *Charles W Morgan* during the operation of trying-out. One man carries the blubber to be made ready for the try-works. Another forks it into the try-works and, while a man stokes the fire, another checks the oil being extracted by boiling.

The Whaling Museum, New Bedford, Massachusetts

The cutting-up procedure was almost the same for the sperm whale and the whalebone whale, but the head of the sperm whale was split into three sections: the lower jaw, which was worthless apart from the teeth which were used for trading or for scrimshaw; the junk, from which came the spermaceti; and the case, which contained the finest grade of sperm oil. When the junk and case were finished with, they were dumped. The head of the whalebone whale was not dissected: the whalebone only was taken out and the rest was discarded.

In 1842, the American fleet accounted for 652 of the world total of 882 ships engaged in whaling. Such was the demand for whale products in America, not only for home consumption but also for export, that the number of whaling ships continued to grow, reaching its peak in 1846 with 735 vessels sailing from American ports. The first bowhead whales (as the Americans called the Greenland right whale) to be killed in the North Pacific were taken off the coast of Kamchatka in 1843 by two New Bedford ships, the *Hercules* and the *Janus*. Four years later this species was found in abundance in the Okhotsk Sea and widely hunted for its great yield of oil and whalebone. In 1848 Captain Roys, commanding the barque *Superior* from Sag Harbor, was the first to sail through the Bering Strait. He reported seeing numerous whales and soon they were being taken regularly in this area.

With the American whaling industry now on the crest of the wave, its vessels were to be found all over the world wherever there were whaling grounds. Though the number of ships had begun to decline in 1846, and continued to fall despite a slight upsurge between 1852 and 1860, there were still a great many occupied in pouring wealth into New England ports, and whale products reached their highest value of $10,730,637 in 1853.

'Whaleman's rendezvous'.

By courtesy of the Seaman's Bank for Savings in the City of New York

By this time, because of improvements to the harpoon, fewer whales were escaping after being struck. The harpoon used by the Americans until the mid nineteenth century was basically the same as that used in the early days of the Spitsbergen whale-fishery, and though in the eighteenth century the British introduced a harpoon with barbs which acted like an anchor once the whale had been hit, the Americans never adopted this weapon. When whales were discovered in the Kodiak Ground in 1835, however, whalemen came into contact with the Eskimos and Indians of the north-west coast of America, and became familiar with an age-old method of whaling to which the Eskimos had added an imaginative refinement. The head of the harpoon had a hole bored through it so that it could swivel on the shaft and, when a strike was made, could be turned at right angles to give a greater hold on the whale. When samples of this weapon were brought back to New England, whalecraft makers along the coast tried to improve their harpoons by using this toggle principle. More than 100 patents were registered within a few years, but one invention so outshone all the others in construction and effectiveness that it was adopted throughout the American whaling industry to the exclusion of any other harpoon. The toggle harpoon designed in 1848 by Lewis Temple, a Negro whalecraft maker in New Bedford, was easy to make yet superbly practical. The toggle adaptation did not impede its powers of penetration, and after entering the whale the toggle turned at right angles giving an extremely high resistance to withdrawal.

In 1852 C C Brand, of Norwich, Connecticut, invented the first successful bomb-lance which was fired from a heavy shoulder gun, but one which became standard, and was largely responsible for killing vast numbers of the bowhead whale, was designed by Eben Pierce of New Bedford. It was used in conjunction with a darting-gun, also invented by Pierce with another New Bedford man Patrick Cunningham. The gun, which had no stock, was fixed at the end of the harpoon pole, with the harpoon attached near the point of the lance, and the complete unit was thrown as a harpoon. When it penetrated the whale, a thin length of metal was forced back, moving a trigger which fired the gun and killed the whale. This procedure shortened the hunt and did away with hand-lancing, so making the actual kill less dangerous.

'The whaler *Dauphin*' by Frank Vining Smith. On 23 February 1821 the Nantucket whale ship *Dauphin*, Captain Zimri Coffin, resued Captain Pollard and Charles Ramsdell of the whaler *Essex* which had been sunk by a sperm whale in the Pacific in November 1820. Their ordeal had taken them to the extremes of human endurance resulting in execution and cannibalism.

By courtesy of the Seaman's Bank for Savings in the City of New York

Attacking the whale.

Author's collection

A fight with a sperm whale.

Author's collection

Whalers and sperm whales.

Author's collection

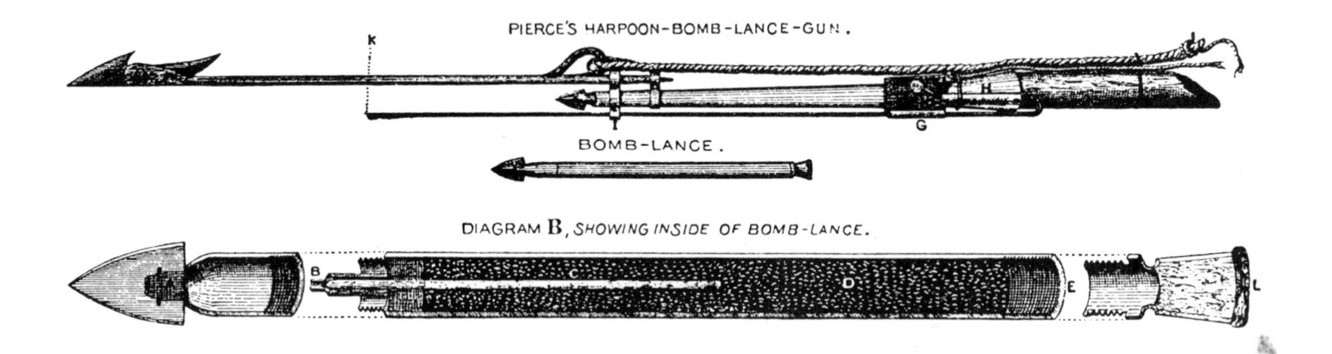

During the boom period of 1835–60 the whaling trade made an important contribution to the American economy, but even while this successful enterprise was at its height, the first indications of things being too good to last could already be glimpsed. The whaling grounds were being extensively overfished resulting in a decline in numbers of whales and so forcing the ships to sail on longer voyages in search of their prey or return home without a full cargo. In time, this situation was to become uneconomic, and capital investment in the trade was considered too great a risk. As factory chimneys made their appearance alongside the whaler's masts, the man with money to invest turned to more lucrative ventures on his doorstep. New Bedford, for instance, the greatest whaling port in the world, whose very existence was owed to the whale, witnessed the beginnings of a cotton industry in 1846, which would in time usurp the whaling trade's prime position in the town's economic structure.

Another contributory factor to the gradual decline of the importance of the New England whaling bases was the opening up of the West, which exercised enormous attraction to the young men of the eastern seaboard, tempting them to turn their backs on the whale trade and seek their fortune in other spheres. The greatest threat to the boom years of American whaling came in 1849 when the word 'Gold!' rang out from the South Fork of the American River in California. Whaling vessels were pressed into service to carry prospectors to the West Coast. Passages were at such a premium that owners anticipated a profitable return by using their whalers as transport ships, but they ran the risk of losing their crews, many of whom signed on merely as a means of getting to the goldfields. It was not unknown for an entire crew, including the captain, to abandon their ship on reaching 'the golden land', the whaling masts towering as mute reminders of the days when salt wind whined in the rigging above a full cargo of oil and whalebone. Deserted ships and rotting equipment represented a loss of thousands of dollars. Despite the difficulties, the whaling trade survived the gold rush and recaptured some of its glory in the 1850s, by which time whaling was being carried out from Californian ports, the *Russell*, late of New Bedford, being the first whaler to be registered at San Francisco, in March 1851.

Pierce's harpoon bomb-lance gun. Made of brass, the weapon was 14in long with an additional 4in provided by the socket and lock-case (G and H). Near the muzzle were two lugs which held the harpoon to which the whale-line was attached. On the other side of the muzzle was a steel rod which extended beyond by about 10in. At the opposite end the rod passed through a tube attached to the socket and hook case as it bent back to the trigger to the lock (J). The whole weapon was mounted at the end of the harpoon-staff. The harpoon was darted by hand. When the harpoon penetrated the whale the rod was driven back so springing the hammer which fired the bomb-lance into the whale. The concussion brought a plunger, held temporarily by a wooden pin in the head of the lance (A) upon the percussion-cap (B) which through the fuse (C) ignited the powder (D) and exploded the bomb.

From 'The Marine Mammals of the Northwestern Coast of North America' by Charles M Scammon

The grey whale follows a migratory route along the west coast of North America, spends the summer in the Arctic and the Okhotsk Sea, and from November to May appears along the Californian coast. There the female was hunted in the lagoons where she bore her young. When exactly the Indians and Eskimos living on the north-west coast of America first started taking these whales, as they passed on their northward and southward journeys, is not known but the white man first took them off Magdalena in 1846.

The Californian grey whale was caught either offshore or in the lagoons. Offshore whaling began from Monterey in 1851 and spread to a number of places along the coast. Shore stations were set up and run on the lines of a whaling ship, using the lay system of pay. The dead whales were towed ashore and dealt with at the whaling station in the same way as New Englanders had done some 150 years before. A number of Portuguese from the Azores were brought over to work in this whale-fishery. Hunting was often carried out from a ship anchored in the lagoon or just outside. Such was the slaughter among this species, especially of pregnant females in the lagoons, that the grey whale was almost exterminated, and only in recent years have they increased in number to a point which recalls the days of the great migrations.

A whaleboat using Greener's harpoon-gun. Similar to a small swivel gun, Greener's gun was of 1½in bore, with a barrel 3ft long, weighing 75lb when stocked, and of considerable accuracy at ranges under 84yds. It was used chiefly in hunting the California grey whale.

From 'The Marine Mammals of the Northwestern Coast of North America' by Charles M Scammon

This whaling industry off the Californian coast made a small but significant contribution to America's whaling prosperity which was not greatly affected by the financial slump in 1857, but the trade faced a major crisis with the outbreak of the Civil War. The Confederate ships *Alabama* and *Shenandoah* played havoc among the American whaling fleet. The *Alabama*, along with some privateers, was particularly active in the Atlantic among the outward- and homeward-bound whalers. Some merchants preferred to leave their ships idle in port rather than refit them and face a loss; others put them into the merchant service, or sold up and abandoned the whaling trade. The United States Government purchased 46 whalers to form what became known as the 'stone fleet'. In 1861 the whalers were loaded with stones and sailed to Charleston, South Carolina, and Savannah, Georgia, where they were sunk at the harbour entrance to deny the ports to the Confederates.

Added to these losses were the number of whalers destroyed by the *Shenandoah* in the Pacific and Bering Strait, which brought the total number of whalers lost during the war to about 90. But the effects were far greater, and through all the variety of reasons the whaling fleet was reduced from 514 to 263 by the beginning of 1866. The discovery of petroleum, which was to become the whale's most formidable rival, in Pennsylvania in 1859 was a further indication that the days of American whaling were numbered. Though the industry continued to operate, with fewer ships, for many years, it was never to regain its former glory.

Australian and Tasmanian whaling enjoyed a boom, alongside that of America, helped by a demand for whale oil in Britain where the heyday of whaling had passed. New whalers were built or old merchant vessels and men-of-war converted to make the most of the boom. Not only did Australian ports, in particular Hobart, benefit from the revenue of their own whalers, but they gained from being ports of call for American whalers.

Australian ships ranged the Pacific covering all the grounds fished by the Americans, but they did not have so long at sea, often cruising for no more than a year. Voyages of shorter duration were made more common by confining the hunt to the Australian sperm whaling grounds which were known as the Middle Grounds (between Sydney and New Zealand), the Northern Grounds (between the Queensland coast and New Caledonia) and the Western Grounds (from Tasmania to the Chatham Islands). As these grounds became overfished the whalers sailed further and further: to the Campbell and Macquarie Islands, Kerguelen, Cape Town, Saldanha Bay, Madagascar, the islands of the South Pacific and eventually to whaling grounds north of the equator.

In the 1850s there were seven whalers out of Hobart making regular voyages to the Bering Strait. They called first at New Zealand and then spent five months among the Pacific islands, working northwards so that they were in Bering Strait for the Arctic season. Their return voyage was by way of the East Indies and, if they still did not have a full ship, they would visit Western Australia. The discovery of gold in Australia affected the whaling industry as men, eager to find a fortune, left the hazards of the whaling trade. This, coupled with the sharp drop in the price of sperm oil, marked the beginning of the end, although there were still 25 whalers owned in Hobart in 1862.

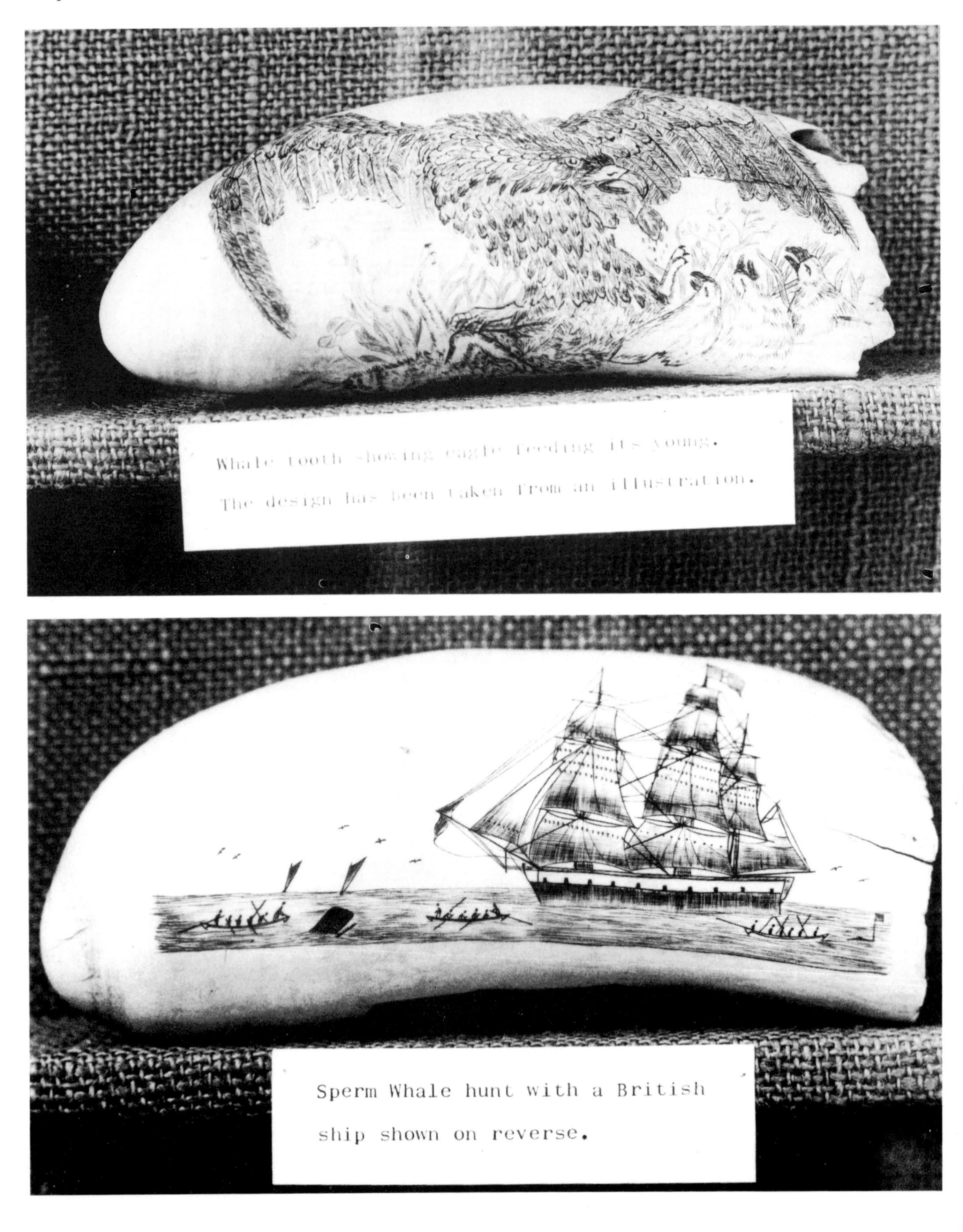

Whale tooth showing eagle feeding its young.
The design has been taken from an illustration.

Sperm Whale hunt with a British
ship shown on reverse.

Long tedious hours of non-activity were spent by some of the sailors in scrimshaw, an art which is almost exclusively the whaleman's, the sperm whale's tooth being particularly suitable for this pastime. The whaleman would depict scenes associated with the sea, or a whaling activity, such as in the sperm whale hunt, a particularly fine example of scrimshaw; or else take this design from an illustration in a magazine, such as that of the exquisite depiction of the eagle feeding its young.

By courtesy of Kingston-upon-Hull Museums and the Seaman's Bank for Savings in the City of New York

The jaw-bones of the whale were used for a number of purposes, one of them being to form the framework of buildings, as in this shed at Whitby.

By courtesy of the Whitby Literary and Philosophical Society

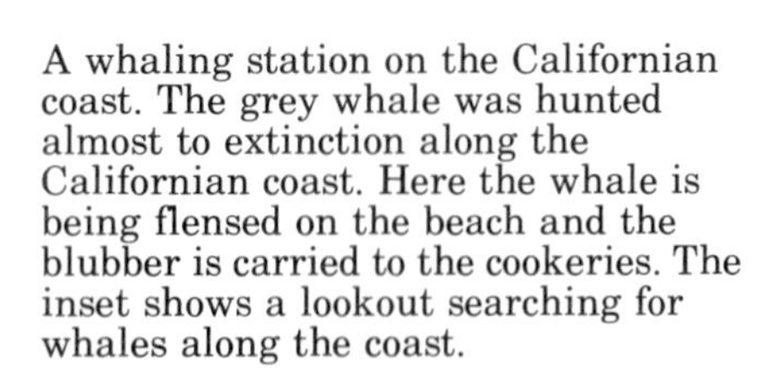

A whaling station on the Californian coast. The grey whale was hunted almost to extinction along the Californian coast. Here the whale is being flensed on the beach and the blubber is carried to the cookeries. The inset shows a lookout searching for whales along the coast.

Author's collection

With the British whaling trade in decline, efforts were made to resume whaling in the South Pacific when the Southern Whale Fishery Company was floated in London with Charles Enderby, son of the famous Samuel Enderby, as chief commissioner. Auckland Island, south of New Zealand, was to be the base for operations. In 1849 the 390-ton *Samuel Enderby* sailed from London with Charles Enderby himself on board. A settlement was made on the island, and in 1851 nine ships were ready for whaling. However, things did not go right, the capital was soon exhausted, and the company was wound up. This was the last venture of the Enderbys and the British in the Southern Whale Fishery. What had once been a great enterprise in British whaling had suffered with the American boom which had won the export market. Vessels were still sent to the seas west of Greenland but only 12 set out in 1852. In 1857 the owners of the famous *Truelove* sent a steam tender to accompany their whaler so that it would not be held up in calm weather. Steam tenders had previously been used in the Arctic by expeditions searching for Sir John Franklin. They were an intermediate stage between the sailing vessels and the auxiliary screw whalers, which really began to come into their own in 1859. Their ability to plough their way through ice was a decided advantage over the sailing ships – steam vessels were full-rigged so that sails could be set in the whaling areas in case the prey was scared off by engine noise.

Attacking right whales at Eden, Australia, in 1840. An artist's impression of the killer whales assisting men to kill the right whale. Twofold Bay was the only known place where killer whales helped in this way. They drove the right whales towards the shore and surrounded them to prevent their escape while the whalers killed them. When the whale was dead the killer whales took it to the bottom and ate the tongue and lips. After the gases brought the whale to the surface about 24 hours later, the whalers towed it ashore for flensing. The whalers of Twofold Bay got to know the killers and gave them such names as Jackson, Typee, Tom, Humpy, Kincher, Albert, Old Ben and Young Ben. In 1878, 27 of them lived in 3 groups known as Stranger's, Cooper's and Hookey's mobs, but all worked together in the whale hunt.

By courtesy of Mr B L Egan, Eden Museum, Australia

The *Diana* was built in Bremen, Germany, in 1840 and made her first voyage to Davis Strait in 1856. The following year she became the first Hull whaler to be provided with steam engines. Captains liked to get local artists to paint a picture of their vessel and the painting may well have been so commissioned.

By courtesy of Kingston-upon-Hull Museums

The coming of the steam whaler did nothing to revive Britain's whaling activities. Whales were becoming scarcer and the demand for their products was on the decline. Hull, which remained the only English port in the trade, sent only two ships, the *Truelove* and the *Diana*, in 1866. The ice conditions were bad that year and at the beginning of September the *Diana* was caught in the ice. By the end of October food was short and parts of the ship had to be broken up for fuel. Scurvy began to take its toll. On 26 December the captain died, and many of the crew owed their survival to the unceasing efforts of the surgeon, Charles Edward Smith. The *Diana* was held in the ice until 12 March 1867 and, to everyone's surprise, sailed into Ronas Voe, in Shetland, on 2 April. The ship was in a dilapidated state; there were ten dead on board and three more were to die later. Thousands of people thronged the docks at Hull to see the *Diana* arrive. There was no cheering. The crowd, who had given her up as lost, greeted her in silence, filled with admiration and respect.

On 11 October the following year the *Diana* arrived off Spurn Point at the mouth of the Humber when a gale drove her on to the Lincolnshire coast at Donna Nook. The crew was saved but the whaler was a complete wreck. The *Truelove* made the last of her 72 voyages that year, before going into other service. So in 1868, Hull's once proud fleet was no more.

It was the year Sven Foyn perfected the harpoon gun which was to revolutionise the ways of whaling.

The *Samuel Enderby*, of 422 tons, shown here as she was in 1834, was named after the owner of the famous whaling firm of Enderby and Sons of London. Charles Enderby, in an attempt to re-establish British interest in the Southern whale-fishery, sailed in the *Samuel Enderby* for Auckland Island, south of New Zealand, to establish a whaling station, but the venture failed.

Author's collection

8 The Birth of Modern Whaling

Sven Foyn, born in Tonsberg, Norway, in 1809, entered the Norwegian merchant service when he was fourteen. Later, when with the sealing fleet, he had the idea of capturing whales by using a gun and bomb harpoon. His experiments proved successful and by 1864 he had a gun which enabled the hunter to hit a whale 50yds away. On entering the whale, the harpoon's barbs opened and broke a glass phial of sulphuric acid, setting off a fuse which exploded the bomb, making a kill almost certain. Thus, the hand-held harpoon and lance were made obsolete.

Realising that the harpoon gun required a sturdier vessel than the whaleboat on which it could be mounted, Foyn had built a steam whalecatcher, *Spes et Fides*, in 1863. The Norwegian's patience and determination paid off in 1868 when he killed 30 rorqual whales which, because of their size, speed and the fact that they sank when killed, were not hunted with the hand-held harpoon from the whaleboat. Foyn overcame the frequent breaking of gear, caused by the weight of the rorqual, by taking the whale-line over a series of springs in the foremast to a winch below. These compensated the tension on the line when the whale struggled and when the dead weight was hauled to the ship. It worked on the principle of the fishing rod which bends under strain, preventing the line from breaking. Foyn's vessel was built with one purpose in mind: to catch rorquals and tow them to the shore for flensing and trying-out; shore stations had therefore to be established, with modern machinery to carry out these operations. So a new whaling industry came into being just when the old one was running down because of the scarcity of the whales which could be killed by the old methods.

A sizeable American fleet still left the New England ports, thriving on the upsurge of a demand for whale products after the Civil War. Part of the fleet was visiting the Arctic regularly, finding whales plentiful in the Bering Strait and beyond. The whalers normally entered the strait in midsummer and stayed until late September or early October, passing south just before ice blocked their exit.

Sven Foyn, a Norwegian, who spent almost his whole life in whaling. His invention of the harpoon-gun, together with the catcher, revolutionised whaling, making it possible to catch rorquals, thus opening the age of modern whaling.

Norsk Sjøfartsmuseum, Oslo

Adverse weather conditions in Bering Strait early in August 1871 were regarded by the captains of 40 American whaling vessels as freakish. Although better weather favoured their decision to stay in the Arctic, its sudden deterioration trapped the greater part of the fleet near Point Belcher, with the ice extending for 80 miles to the south. The decision was taken to abandon ship, and 1219 men, women and children made the journey in open boats, sometimes having to drag them across the ice, to the safety of five American whalers, one from Honolulu and one from Australia. A total of 34 American vessels were lost.

Author's collection

A modern harpoon-gun with harpoon loaded. The forerunner of this type of gun was invented by Sven Foyn in 1864. The modern weapon hurls a 250lb explosive harpoon at the whale.

Institute of Oceanographic Sciences

This early type of harpoon-gun was fitted on a swivel mounting in the bow of the whaleboat. This particular gun was made by G Wallis of Hull in 1816 and was in use on the whaler *Phoenix* of Whitby.

By courtesy of the Whitby Literary and Philosophical Society

Forty vessels had passed through the Bering Strait by 30 June 1871 in excellent weather conditions which continued through into August. On 11 August there was a sudden deterioration and ice formed rapidly, forcing the ships into a narrow strip of water near Point Belcher. Better weather came on 25 August and a passage opened to the fleet. The captains regarded the adverse conditions as freakish and decided to stay in the Arctic, continue whaling and leave at the usual time. This decision proved fatal. The unusual weather returned on 29 August and trapped the fleet. An expedition, sent to explore the possibilities of escaping, reported that the ice extended for 80 miles south and there was no chance of freeing the ships. There were two vessels beyond the ice and five trapped near the edge but with every likelihood of reaching open water. Partly because there were women and children on several ships, the decision was made to risk the hazardous journey south rather than face possible starvation. 1219 men, women and children undertook the journey in open boats using any available strip of water but at times having to drag them over the rugged terrain while enduring icy wind and ferocious gales. All reached the seven ships, five of which were from American ports, one from Honolulu and one from Sydney, New South Wales. Thirty-four vessels were lost and the total financial loss has been estimated at over $2,000,000.

The *Spes et Fides*, the whalecatcher invented by Sven Foyn so that his harpoon-gun could be used effectively.

Norsk Sjøfartsmuseum, Oslo

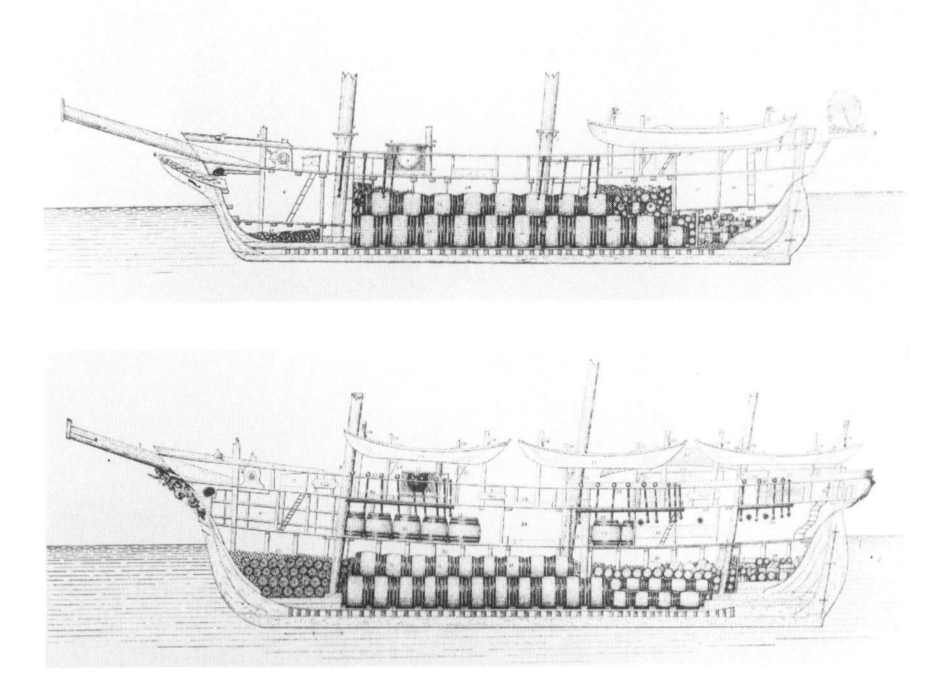

Sectional drawing of the schooner *Amelia* (95 tons). This vessel was engaged in whaling out of New Bedford from 1876 until lost in 1881 at Abrothas Island, Brazil.

Author's collection

Sectional drawing of the barque *Alice Knowles* (303 tons), which sailed out of New Bedford from 1879 until 1889 when she changed her port to San Francisco, reverting to New Bedford in 1908. During the voyage of 1910–1913 the crew mutinied and refused to lower the boats to pursue whales. She left New Bedford in 1914 to hunt whales in the Atlantic and was lost in a hurricane on 2 September 1917.

Author's collection

This disaster, together with the loss of 12 ships in the Arctic in 1876, 5 in 1888 and 8 in 1897, came at a time when there was less incentive to replace the ships. In the last quarter of the nineteenth century more lucrative fields attracted investment in a growing nation and there was no stimulus to continue in, or embark on, a trade which was suffering as petroleum became more widely used as a means of illumination and a lubricant. The whaler sailed in ever-decreasing numbers. Nantucket's last vessel to arrive with a cargo of oil was the brig *Eunice H Adams*, on 16 June 1870. In 1880 only 9 ports were engaged in whaling and by 1896 only 77 ships sailed to hunt the whale.

As with British whaling the decline was only partially halted by the introduction of the steam whaler. The Americans were slow to adapt steam to their whalers, being 23 years behind the British who first used it in 1857. The steam whaler cost over three times as much as the sailer and had to be manned by twice as many men. Coal was scarce, making running costs high. To save the expense of the long voyage back to California and then back to the Arctic the following season, many of the American steam whalers wintered in the Arctic, sending their produce back by sailers.

The combined use of sail and steam contributed to San Francisco's rise as a whaling port, as did the transcontinental railway. The Rockies and Sierra Nevada were no longer formidable barriers isolating San Francisco from the east. Whale products landed there were transported by rail to eastern towns, saving the ship the long and hazardous voyage round Cape Horn. In 1893, 30 whalers were registered in San Francisco but with many more whalers using the port as headquarters it was becoming the leading whaling port in the country. In 1906 it was one of only three ports left in the whaling trade; New Bedford had 24 vessels, San Francisco 14 and Provincetown 3, although that year Norwich had one brig whaling, the first time for 27 years.

Right Sectional drawing of the *Hope*, built by Alexander Hall and Co of Aberdeen, Scotland for Captain John Gray of Peterhead in 1871. This and his brother's vessel, the *Eclipse*, were built along new lines and proved highly successful. The *Hope* was sold to Newfoundland in 1891.

From 'Souvenirs de Marine' by Amiral Paris

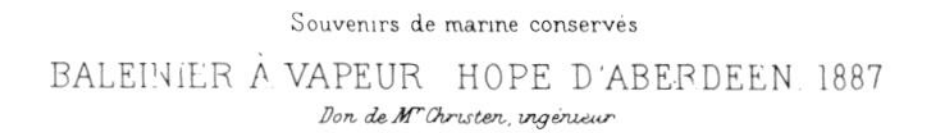

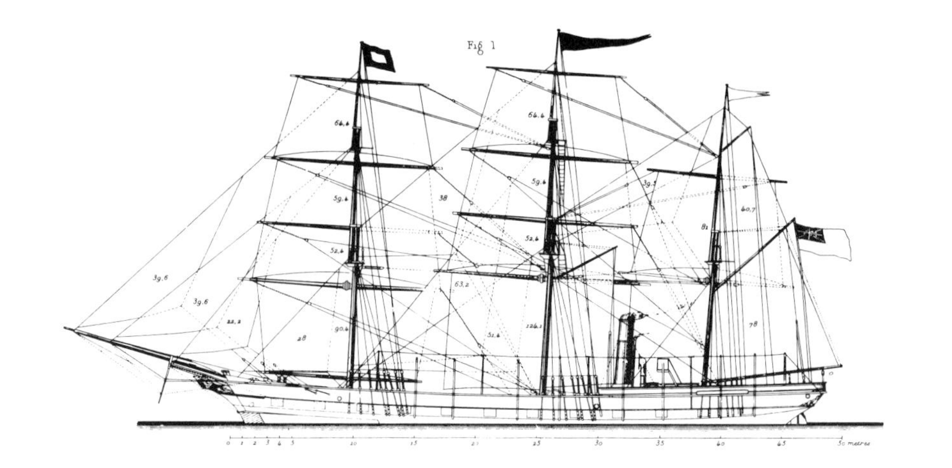

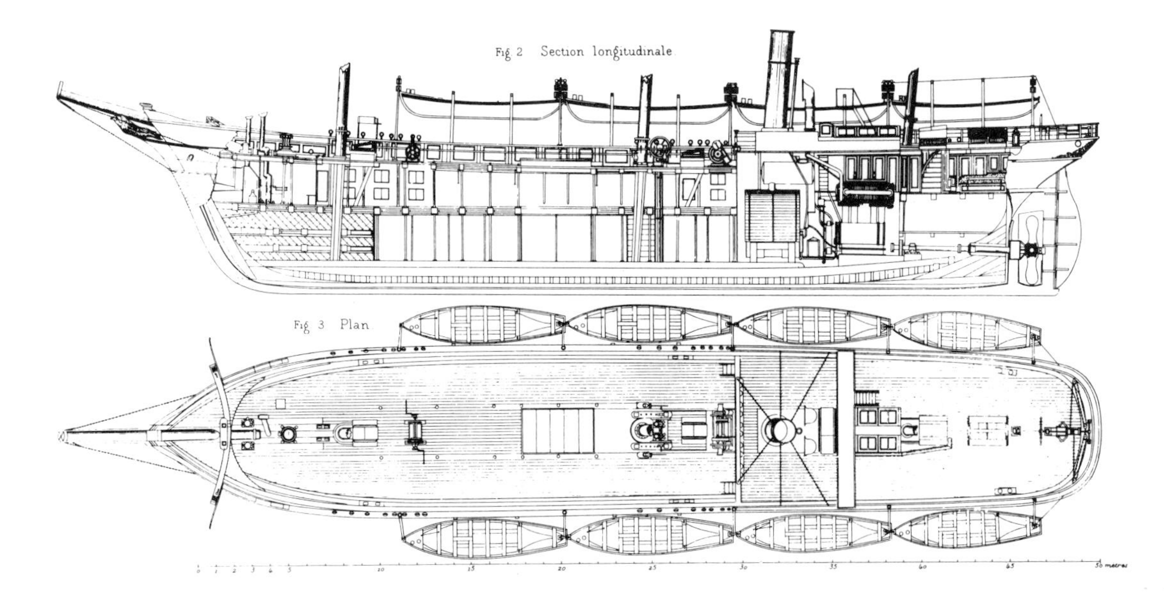

Le principe le temps est de l'argent qui a fait disparaître presque partout les voiles pour les remplacer par une machine plus coûteuse de création et surtout d'emploi, se trouve avoir étendu son influence jusque sur la pêche la plus grandiose et la plus chanceuse, qui expose un tel navire à perdre les avantages de son moteur, si le hasard ne lui fait pas rencontrer de baleines. Il en est pourtant ainsi et cette collection consacrée surtout au passé doit de la reconnaissance à Monsieur Christen, ingénieur d'Aberdeen, pour lui avoir communiqué les plans et une note sur cette nouveauté si intéressante. Il a obtenu ces documents du constructeur Mr Hall.

La manière d'agir des baleiniers à vapeur diffère de celle des autres, en ce qu'ils s'éloignent moins, reviennent à chaque saison et par suite ils évitent de faire fondre la graisse à bord, mais ils la placent dans des caisses en tôle au nombre de 38 sur le Hope. Elles sont dans le fond de la cale, comme on le voit sur la section et contiennent une tonne de graisse qui, fondue à terre donne 312

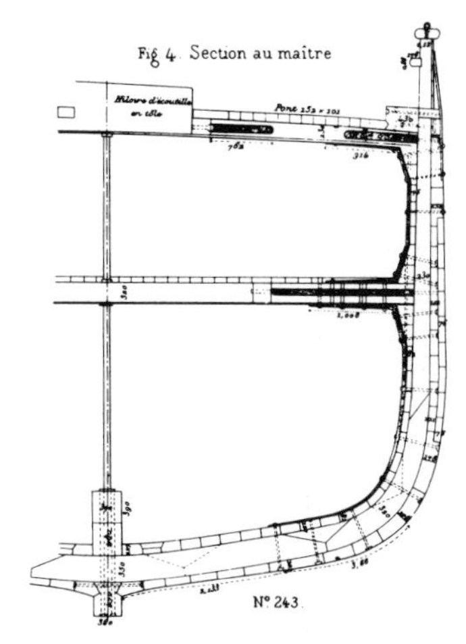

gallons ou 1416 litres d'huile.

La construction de ces navires est très soignée très solide, comme on peut en juger par la section. Le fer y est employé à la liaison comme le montrent les parties teintées. Les hiloires des écoutilles sont garnies de tôle.

Les formes et les proportions sont celles d'un clipper de marche. La machine est à pilon, sa force de 70 chevaux, les cylindres ont 0m586 et 1m111, le diamètre de l'hélice est 3m80 d'après le dessin. Les figures montrent bien la disposition du pont et de l'intérieur, tandis que l'élévation du travers expose la mâture et le gréement qui, pour le dormant est en fer.

Les dimensions principales sont longueur entre perpendiculaires 45m72, largeur extrême 8m534, rapport 1 à 5,36, creux 5m181, tonnage 307tx anglais. La section porte les dimensions des différentes pièces. La surface de voilure est 692m et les voiles d'étai 192, maître couple 34m5. Rapport 1 à 20

The British whaling industry was also a shadow of its former self and only the Scottish ports of Peterhead and Dundee really kept it going through the latter part of the nineteenth century and into the twentieth. In 1861 Peterhead's fleet numbered 21 vessels and Dundee's 8. Dundee took over the leading role in the 1870s, though two of the best whalers operated from Peterhead. These were the *Eclipse* and the *Hope*, auxiliary steam whalers captained by David Gray and his brother John, members of a well-known whaling family and so skilful that throughout this period they continually made a profit.

With the traditional hunting grounds in the north almost depleted of whale stocks, whaling men, among them the Gray brothers, turned their eyes towards the schools reported by explorers of the far south. The Grays proposed a joint venture with some friends to establish a whale-fishery in the Antarctic. They expected to hunt whales of a species similar to the Greenland whale which they had been hunting in the Arctic. In spite of their confident report they failed to raise the necessary capital and almost twenty years elapsed before the whole question was examined again.

As the whale stocks in the seas around Greenland diminished, the Scottish whalers started to take the smaller bottlenose whale. Unfortunately for the Scottish whalers the Norwegian sealers also started to hunt the bottlenose. Overkilling resulted and the price of bottlenose oil dropped from £90 a ton to £18, making it no longer profitable to the Scottish whalers, though remaining so to the Norwegians who used smaller, less expensive vessels. Their small 30–50 ton schooners were armed with up to six guns. If the whales were out of range of the ship then boats were sent in pursuit. Two men rowed, one steered and one fired the small gun mounted on the bow. The harpoon was only 3ft long but strongly barbed. In 1885 the Norwegians killed 1300 whales from 30 ships, and they steadily increased this to 70 vessels which took 3000 whales in 1891, ignoring the drastic effect this had on stocks.

The *Antarctic* in the far south. H J Bull, who worked in the mercantile house of Trapp, Blair and Company in Melbourne, Australia, was convinced that there were plenty of right whales in the Antarctic. Australians were not interested so he visited Sven Foyn, then 84, in Norway and they put a retired steam whaler of 226 tons, which had been built in 1872, into service and renamed her the *Antarctic*. She sailed from Melbourne on 12 April 1894. Equipped to hunt only right whales, the expedition met no success, but it left Bull convinced that a successful whaling industry could be mounted from shore stations in the Antarctic. He failed, however, to arouse sufficient interest.

Author's collection

The *Arctic* in the Arctic. The auxiliary steam whaler *Arctic* of Dundee (439 tons) was built in 1867 for the whaling trade. She made seven whaling voyages and was lost in Prince Regent Inlet in 1874.

Author's collection

The possibilities of an Antarctic whale fishery had come up again and in 1892 the Tay Whale Fishing Company of Dundee despatched four ships, the *Balaena, Active, Diana* and *Polar Star* to the Antarctic equipped to take right whales. None was seen, but the men returned with reports of seeing numerous rorquals. Three Scottish naturalists, led by Dr W S Bruce, who were on this expedition, tried to arouse interest in establishing a whaling industry, using the Norwegian methods, in the Antarctic. In spite of the lure of lucrative returns, they were unsuccessful in raising sufficient capital. Similarly, H J Bull, who lived in Melbourne, Australia, and Carl Anton Larsen, a Norwegian, both failed to arouse enthusiasm for what they considered would be lucrative whaling projects in the Antarctic. A new century was to arrive before the Antarctic was to hear the boom of the harpoon gun heralding the destruction of the large stocks of rorquals.

As Sven Foyn's method of whaling was establishing itself in the northern hemisphere the old methods were petering out. The right whales were scarce and the chances of making a profit from whaling alone by the old system were slight. By 1896 Dundee was the only British port left in the trade. Gradually the fleet diminished until in 1913 only two vessels, the *Morning* and the *Balaena*, sailed to the Arctic, but they returned without success. The outbreak of World War I saw the end of the old whaling industry.

The *Arctic* hunting the whale.

Author's collection

Sven Foyn's successful combination of harpoon gun and catcher meant that the new world of rorqual hunting progressed by leaps and bounds. In 1871 whale flesh was successfully turned into fertiliser, which made the killing of the lower oil-yielding fin whales worthwhile. By 1887 there were 20 companies operating 35 catchers, and the Norwegians had become the leading whaling nation. But the slaughter was reaching such proportions that already the number of whales within suitable towing distance of the shore stations was declining. Five companies moved to Iceland where the biggest shore station was set up at Onundarfjordur in 1889, and in the next eleven years 1296 blue whales were killed from that station alone. Stations were built in the Faeroes and the Hebrides, and in 1897 the first shore station began operating in Newfoundland. Within eight years there were 18 such stations there.

The Norwegians had started to span the world, a process which was speeded up by conflict with fishermen at home, who claimed that whales drove fish closer to the shore and were therefore essential to the well-being of their trade. These fishermen saw the killing of whales on such a scale as a threat to their livelihood. Matters came to a head with riots in Mehavn and the Norwegian Government brought in legislation prohibiting the pursuit, killing and landing of whales in the territorial waters of Norland, Tromso and Finmark for ten years from 1 February 1904. Anticipating the result of the enquiry, two Norwegian companies commenced operations on Ronas Voe in Shetland in 1903. The following year shore stations were also established at Colla Firth and in the Hebrides by the Norwegians and in the Hebrides and in Shetland by the Danes.

'The *Diana* trapped in the ice' by T Widdas, 1867. This scene represents the day of 2 December 1866, taken from a sketch by Charles Edward Smith, surgeon on the *Diana*. This ship was eventually freed on 17 March 1867, her crew having suffered much hardship during the Arctic winter. In 1868 the *Diana* ran ashore at Donna Nook on the Lincolnshire coast and was lost.

By courtesy of Kingston-upon-Hull Museums

At first there were no restrictions placed on foreigners whaling from Scotland, but when complaints came in from local herring-fishing interest, the Secretary of State for Scotland appointed a Committee of Inquiry to look into the matter. The committee decided that unrestricted whaling could be a menace to herring-fishing but there was no case for its total abolition. Regulations were laid down followed by the Whale Fisheries (Scotland) Act of 1907, which empowered the Scottish Fishery Board to exercise general control over the industry.

In 1905 catchers operated for the first time from a shore station on the Pacific coast of America at Sechart on Vancouver Island. Others were established in the Aleutian Islands, Alaska, Mexico and south of the equator.

Modern whaling came to Japan through the Russians. In 1891 Tsar Nicholas II visited Japan and on his return journey to Russia saw whales in sufficient numbers in the Sea of Japan to arouse enthusiasm to form the Russian Pacific Whaling Company. It began operations in 1898 and, using Norwegian equipment and methods, surprised the Japanese net-whalers with the number of whales it could kill and deal with compared to their methods. Jyuro Oka ordered a whalecatcher, called it the *Chosyu Maru*, and went to Norway to study the new methods at first hand. On his return he founded the Japan Ocean Whaling Company basing it at Senazaki, a net-whaling station, on the coast of the Sea of Japan. Fifteen whales were killed in 1899, 42 the following year and 60 in 1901. Unfortunately the *Chosyu Maru* was wrecked that year, but, undeterred, Oka chartered three ships, the *Orca* from England and the *Rex* and *Regina II* from Norway, so that no time was lost in keeping the company going.

A model of the whale-ship *Viola*. Of 139 tons, the *Viola* made her first whaling voyage in the Atlantic, leaving New Bedford in 1910 and returning two years later. On this voyage she took 150lb of ambergris, worth $30,000. After leaving New Bedford on 5 September 1917 she was lost with all hands.

By courtesy of the Seaman's Bank for Savings in the City of New York

Jyuro Oka, with Japanese keenness and adaptability, had shown that the new methods of whaling could be a boom to Japan and the last time nets were used for whaling was in 1909 in south-west Honshu. Several other companies were formed and Japanese whaling expanded not only around the coast of Japan but also in Korea, on the Kurile Islands and in Taiwan. During their period of expansion the Japanese engaged Norwegian gunners and captains to teach them. Throughout the world the general practice was for the Norwegian gunner to be in complete control, but eventually the Japanese Government intervened, in the case of their own whaling enterprises, and forbade anyone but a Japanese to be in command of the ship. Rather than lose the expertise of the Norwegians some Japanese operators got round the regulations by handing over to the Norwegian gunner once outside Japanese waters.

Japanese whaling thrived and has continued to do so with war causing the only serious set-backs. The shore bases were eventually grouped into two companies, Taiyo Gyogyo and the Nippon Suisan Company, both of which still operate today. The Russo-Japanese War of 1904–5 put an end to the Russian interest in whaling, but the Japanese continued to expand the industry because of the demand for whalemeat for human consumption.

The Norwegian type of shore station and their method of dealing with dead whale were in general use throughout the world, with the exception of Japan. All shore stations were generally established in a sheltered cove or bay. The Norwegian method necessitated a flensing slip on which the whale was positioned for the operation of cutting-up. After incisions had been made the blubber was stripped off by a hook attached to a cable which was pulled by a winch. The blubber was then fed into slicing machines from which it was put into vats to be tried out. Once all the blubber was taken off, the carcase was split open and the heart, intestines and lungs were extracted.

The carcase was then put on to the carcase platform which was slightly higher than and at right angles to the flensing slip. Here the rest of the flesh was pulled off, cut into chunks and boiled to extract more oil. Similarly the bones were boiled for yet more oil, after which they were crushed and made into fertiliser. The remains of the boiled flesh were dried and turned into a fine guano. The blood was carefully preserved during the operation and turned into a dry fertiliser. Even the water in which the blubber had been boiled was kept and turned into a glue. The baleen was cleaned and dried, so that practically every part of the dead whale was used.

The Japanese method of dealing with the dead whale resembled that used by the whalers at sea. At the end of the wooden wharf two upright poles were joined by a cross-piece on which were placed the necessary pulleys, thus resembling the gear attached to the masthead of a ship prepared to flense at sea. Wire cables were fixed just in front of the whale's flukes and the whale was pulled upwards out of the water. Two cutters in a sampan beside the whale split the body in front of the dorsal fin. The piece attached to the pulleys was drawn on to the wharf and the work of cutting it up, which had started as it was being drawn out of the water, continued when it was lowered on to the wharf. Cutters incised the part remaining in the water and, in the fashion of the high seas whalers, a hook was inserted into the blubber and, by means of the

This detail of the *Viola* model shows clearly the cutting stage, the try-works and the position of the whaleboats.

By courtesy of the Seaman's Bank for Savings in the City of New York

A whale coming up after 24 hours when the killer whales had eaten the cheeks and tongue – Eden, Australia.

By courtesy of Mr B L Egan, Eden Museum, Australia

winch, blanket pieces were removed from the body as it rolled over in the water. These blanket pieces were passed to the people on the wharf who cut them into blocks. All the pieces of flesh and blubber were carried on carts to the washing vats.

This work went on continuously as long as there was a whale to flense, fires being used for illumination if night work was necessary. Nor did the work stop if the weather was poor. As the whale formed an important part of the Japanese food supply, there was an urgency to get the whale meat and blubber to the towns and cities, where it was sold in the markets or from door to door. In some parts canning factories were developed so that in summer, when the heat made the transportation of fresh meat over long distances impossible, the flesh was cooked and packed in tins. The head which was severed from the body was dealt with separately, and though the whalebone of the fin whale was not of great commercial value, being short and stiff, the Japanese made sandals and numerous types of trinkets from it.

With disregard for the effect on stocks, killing from the shore stations across the northern hemisphere went on unabated. In 1902, 1305 whales were killed by 30 catchers sailing from Iceland. Ten years later only 15 whales were taken. Drastic measures had to be imposed in order to try to save the situation around the Iceland coast for the future. In 1915 whaling from the island was made unlawful, a ban which was to go on for twenty years. Realising that the days of lucrative returns from their shore stations in the north were numbered, the Norwegians moved south. Stations were established in South America at Valdivia and San Pedro in Chile and on the coast of Brazil.

Migrating humpbacks had been observed off the African coast, so the Norwegians sent an old catcher, *Neptune*, and a floating factory, *Vale*, to Saldanha Bay, on the south-east coast of Africa, in 1909. Reports were good and in 1910 shore stations were built at Durban and Saldanha Bay. By 1913, 25 stations were spread along the African coast

Flensing the whale – Eden, Australia.

By courtesy of Mr B L Egan, Eden Museum, Australia

from the Congo to Angoche in East Africa. A large amount of capital had been sunk in this venture and it needed a large return in oil and other products to make a profit. This, coupled with the rivalry between stations (often resulting in foolish competition), meant prodigious killing of the migrating humpbacks which followed the line of the coast, never more than twelve miles out. So these whales were an easy target and were attacked on their way to and from the breeding grounds, when they should have been left alone so that the continuation of stocks could be assured. But man's greed pushed the slaughter on and in 1912 it was estimated that if the butchery went on at the same rate the whales along the African coast would be exterminated within six years. No attention was paid as profits were relentlessly pursued.

The Norwegians also looked to the waters of the old Australian whalers. A factory ship was off Tasmania in 1911 and, having no success, moved on to New Zealand, but still without much luck. The real effort came the following year when the Norwegians investigated reports of migrating humpbacks off the coast of Western Australia. The Spermacet Company of Larvik floated the Western Australian Whaling Company and, with the Australians eager to develop this section of the coast, gained a licence giving them exclusive rights for seven years, and the first expedition moved in in 1912. Disputes arose when other companies appeared, but the number of humpbacks was not as large as expected and by 1917 all the whaling companies had left the coast of Western Australia.

While all this had been going on the Antarctic was being opened up for the greatest whaling enterprise – and the biggest slaughter – of all time. Captain Larsen, who had been disappointed at not being able to interest anyone in Antarctic whaling, returned to that world of ice as captain of a Swedish exploration ship. The ship was lost in the ice but fortunately the crew and expedition members were saved by an Argentinian gunboat, the *Uruguay*, and taken to Buenos Aires. While here

Flensing – Twofold Bay, Australia. The whale was flensed in the water. A rope attached to a man-powered capstan peeled the blubber while the flensers cut it away with flensing-knives and boat-spades.

By courtesy of Mr B L Egan, Eden Museum, Australia

Larsen put his idea for an Antarctic whaling industry to a group of Argentinian businessmen. They were enthusiastic and formed the Compania Argentina de Pesca with Larsen as its whaling manager. He returned to Norway where he fitted out two sailing ships and a steam whalecatcher. In December 1904 a delighted Larsen sailed into the bay he had chosen in South Georgia as an ideal site for a shore station. He called it Gryt Vik, or Pot Bay, after finding round iron blubber pots left on the beaches by sealers. The definite article was added later and it became known as Grytviken, The Pot Bay. Humpbacks and right whales were plentiful and the company had a good first season, but better years were to follow. South Georgia was a British possession and the whaling station had been established there without the knowledge of the Government of the Falkland Islands who administered the territory. The government made it compulsory to have a licence to establish a whaling station and to pay royalties on the catch. Larsen's company was granted the necessary licences for 21 years from January 1906.

In 1903 the Norwegian, C Christensen, was first to use an oil-boiling installation on board a steamer. The following year the *Admiralen* was fitted out with open boilers and used for whaling off Spitsbergen. The floating factory system was born and the company, Messrs Lars Christensen and Sons, of Sandefjord, applied for a licence to whale at the South Shetlands. The *Admiralen*, with two catchers, had a successful season in 1905–6. The whale was flensed alongside the ship and tried out on board. The oil was stored in the ship ready for transport home. The vessel could operate independent of shore bases, which meant that, while it required a sheltered bay for an anchorage, it could move as necessary and the whale could be hunted in places where the coast was unsuitable for shore stations. South Georgia became the new centre of the new whaling era. By 1909 another six companies had successfully applied for licences and leases and the whale in South Georgian waters was doomed.

Flensing – Kiah River, Eden, Australia.

By courtesy of Mr B L Egan, Eden Museum, Australia

The humpback, because it was plentiful and found close to the coast, was the main object of attack in the early days of the industry. In 1910–11 6197 humpbacks were killed out of a total of 6520 whales taken. From then on the number of humpbacks declined and the catchers switched their attention more and more to the fin and blue whales.

Whaling spread to other areas of the Falkland Islands Dependencies and man killed indiscriminately without thought for his victim or for the future of the whaling industry. He imagined there would always be plenty of whales for he knew little of the lengthy natural process of stock replacement. It was the same during World War I, which brought a great demand for oil, fertiliser and especially for glycerine for use in explosives. The slaughter went on unabated and declined during two of the war years only because of shipping losses.

However, some concern was beginning to filter through. The British stated that females with calves must not be killed and required each company to send to the Natural History Department of the British Museum a record of every whale killed, detailing the date of capture, species, sex, length and, if pregnant, the length of the foetus. This was a start, but there was little to restrict the slaughter. There was no regulation on the number of whales which could be killed, and during the chase it was difficult to tell if the whale *was* a female with calf; even if it was, who would know, apart from the crew of the chaser?

The immense amount of data collected by Major Barrett-Hamilton of the Natural History Department of the British Museum when he visited South Georgia in 1913 formed the basis for work which followed after the war. In 1917 the Government of the Falkland Islands levied a tax on each barrel of oil to provide a fund for researching the natural history of whales and out of this the Discovery Committee was financed after the war.

WANDERER

9 The Antarctic Boom

After World War I the Americans tried to revive the whaling industry by fitting out some of the old vessels, but these could not compete with the modern whalecatcher. On 25 August 1924 the *Wanderer* of New Bedford set sail on a whaling voyage and anchored off Mischaum Point for the night. During the early hours of 26 August she was driven ashore and wrecked on Cuttyhunk Island. She was the last of the American square-riggers. After the *John R Manta* set out from New Bedford in 1925 no more whaling ships left port.

The only surviving whaler from America's golden days is the *Charles W Morgan*, built in 1841. She undertook 37 voyages under 20 captains before returning from her last whale hunt in 1921. After taking part in a number of films, she lay derelict in Fairhaven but was rescued by a group of New Bedford men. Battered by a hurricane in 1938, she again became a hulk, but in 1941, thanks to the Marine Historical Association, Mystic, Conneticut, the ship came to her final resting place in the Marine Museum there, serving as a permanent reminder of the days when wooden ships set sail to hunt the whale.

America re-entered the whaling trade in the 1930s, operating from shore stations along the Pacific coast of North America and also from chartered factory ships in the Antarctic and off the Australian coast. These were all small operations, as was the new shore station built at Fields Landing in California in 1939 which was used until 1949. In 1935 the ban on whaling from Iceland was lifted and a station established at Talknafjordur operated until World War II. The Japanese worked a successful whaling industry from shore bases throughout their empire but their most lucrative days were to be spent in the Antarctic.

The Australians failed to utilise the shore stations established by the Norwegians before 1917, and did not capitalise on the knowledge they had gained from the experts. A company formed in 1921 failed and was only saved by the Norwegian Bay Whaling Company which gradually put the Australian concern on a profitable footing. Even while this

The wreck of the *Wanderer*. This vessel was the last American square-rigged whaler in service. Putting to sea on 25 August 1924 from New Bedford, she was wrecked on Cutty Hunk Island on the following day.

The Whaling Museum, New Bedford, Massachusetts

was happening a short-sighted government was trying to impose a royalty on every whale caught, so the Norwegians thought about leaving. Before a final decision was reached the Norwegians no longer needed the shore station, for the pelagic whaler had arrived and there was more profit in the open seas of the Antarctic. From then on the Australians merely debated the possibilities of developing a whaling industry, and missed the advantage of their nearness to the Antarctic whaling grounds.

Whaling from Natal has continued since its inception. In the 1930s two companies, the Union Whaling Company and one run by Lever Brothers, were working out of Durban. Each had a shore station standing on a low sandy cliff facing the Indian Ocean a short distance out of town. The plan (flensing area) sloped away from the sea instead of into it as was usual; the waste drained into gutters and was piped away to the sea. The position and height of the plan led to another peculiarity: the dead whale was brought to a slipway further round the Bluff where it was lifted on to trucks and taken to the whaling station by rail. The train ran alongside the plan where winches pulled the whale from the trucks, to be flensed by Zulus supervised by Norwegians.

'The *Wanderer* on the whaling grounds' by Gordon Grant, 1946.

By courtesy of the Seaman's Bank for Savings in the City of New York

A model of the *Charles W Morgan*.

By courtesy of the Seaman's Bank for Savings in the City of New York

The Norwegians were involved in practically every whaling enterprise in the world either through their own companies or as employees of other countries. They were the experts. While Antarctic whaling was their main sphere they did not neglect any other promising opening. In the early 1930s they started a specialised whaling industry in the seas off Norway and Spitsbergen and between Norway, Iceland and Greenland. Only World War II interrupted this activity. Small whales were hunted from highly manoeuvrable ships of between 40 and 120ft, designed so that the minke, bottlenose and pilot whales could be cut up on board.

South Georgia
1. Inhospitable Bay
2. Possession Bay
3. Stromness Bay
4. Husvik
5. Cumberland Bay
6. Grytviken

The main area of whaling operations between the two wars was the Antarctic, centred, at first, on the Falkland Islands Dependencies, chiefly South Georgia. The routine generally followed at these shore stations began when the transport ship arrived with supplies and men to operate the catchers and their base. The catchers were in action as soon as possible and, once whales started to arrive for flensing, there was no respite for those at the station. Meanwhile the transport ship's tanks, which had contained coal for the shore station, were thoroughly cleaned. If the hunting was good a cargo of oil was ready by Christmas. This was taken to the north and the transport aimed to be back in the Antarctic during March. After unloading its second cargo of coal it was ready with its final cargo of oil by the end of the season. A few men were left behind to overhaul the machinery and catchers for the next season.

As whales became scarce around the coasts due to overkilling men began to look for new areas in which to keep the industry alive. Carl Anton Larsen, who had retired to his Norwegian farm in 1914, remembered the untouched whales far out to sea and particularly those reported by the explorers of the Ross Sea. The urge to return to the Antarctic was so strong that in spite of being sixty, he decided to equip an expedition to hunt the whales in the Ross Sea. At Liverpool he purchased a steel cargo steamer, the *Mahronda*, a roomy vessel of 13,000 tons with ample deck space and strong engines. He had it converted to a floating factory in Sandefjord. The hold was turned into 22 tanks capable of holding about 60,000 barrels of oil. Boiling machinery and a new German machine for extracting oil from bone were installed. The ship's equipment was the very latest and included the most powerful wireless set in any merchant vessel. Larsen changed her name to *Sir James Clark Ross* in memory of the discoverer of the sea which he hoped would become the cornerstone of a new whaling industry.

A Durban whaling station from the air. The whaling stations at Durban were established on a long stretch of sandy beach, washed by the Indian Ocean, to the south of the town's beautiful landlocked harbour.

Institute of Oceanographic Sciences

A sperm whale on a rail truck at a Durban whaling station. The railway was a unique feature of whaling at Durban. It ran from the slipway, close to the harbour entrance, about two miles to the whaling stations. The carcase was winched on to low-slung wagons specially built for the purpose. At the whaling station, the plan (the area on which the carcase was dealt with) was built like a station platform alongside the track.

Institute of Oceanographic Sciences

A fin whale on a rail truck at a whaling station, Durban.

Institute of Oceanographic Sciences

Sperm whale on a rail truck at a whaling station, Durban.

Institute of Oceanographic Sciences

A sperm whale on a rail truck at a Durban whaling station.

Institute of Oceanographic Sciences

Flensing a humpback whale, Durban.

Institute of Oceanographic Sciences

Flensing a sperm whale, Durban.

Institute of Oceanographic Sciences

Saldanha Bay whaling station, Cape Province, South Africa, about 1926. Saldanha Bay whaling station was established in 1910 by Consul Johan Bryde after whom the Bryde's whale was named. It operated until its closure in 1930.

Institute of Oceanographic Sciences

Fin whale, Saldanha Bay, about 1926.

Institute of Oceanographic Sciences

No matter how efficient he made the floating factory it was useless unless served by a fleet of catchers large enough to complete the long voyage to Antarctic, and powerful enough to cope with the ice pack and severe weather conditions. Larsen found two locally in Sandefjord and three in Seattle, USA. These three were American-built to Norwegian design and had operated off the Alaskan coast for eleven years. Most of the crew Larsen picked were experienced men who had seen fifteen to twenty years in whaling and many had served in South Georgia. The British Government granted Larsen a lease of the Ross Sea for five years but laid down stringent conditions, and no other whaler was allowed to operate in the area.

The ships sailed in 1923 and met at Hobart, Tasmania. In spite of severe weather conditions the expedition successfully penetrated the icefield, entered the Ross Sea and, on Christmas Eve, reached the Bay of Whales. On Boxing Day, Captain Iversen, on the catcher *Star II*, killed a blue whale – the first to be taken in the Ross Sea – but the expedition was faced with a problem: blue whales could not be lifted on board, as the tackle was only strong enough to lift humpbacks. The blues would have to be flensed alongside the ship; this required a sheltered place, but the Bay of Whales was frozen in. The expedition went to Discovery Inlet, the only other harbour in the ice barrier. This was open and, in spite of fogs, blizzards, ice and temperatures which sometimes dropped to 55°F below freezing point, the expedition operated from 31 December 1923 to 7 March 1924.

Having to work from Discovery Inlet instead of in the open sea meant that the catchers had to tow their kills over greater distances, taking up valuable hunting time, and only 17,500 barrels of oil were processed instead of the expected 40,000 to 60,000. In this respect the expedition was something of a failure, but a vast amount of experience was gained and the German boilers, which had been tried only once before, proved successful.

King Edward Cove and the Pesca Company's whaling station, Grytviken. Captain Carl Anton Larsen, seeing the possibility of exploiting the whales around South Georgia, persuaded some businessmen to form the Compania Argentina de Pesca in Buenos Aires. He chose a site in King Edward Cove for his whaling station and called it Grytviken.

Institute of Oceanographic Sciences

Blue whale (22.3m), Grytviken 1925.

Institute of Oceanographic Sciences

Fin whale (21.3m), Grytviken 1925.

Institute of Oceanographic Sciences

Blue whale (27.2m) on a flensing platform, Grytviken 1925.

Institute of Oceanographic Sciences

Flensing a humpback whale (14.1m), Grytviken 1926.

Institute of Oceanographic Sciences

Male sperm whale, Grytviken 1926.

Institute of Oceanographic Sciences

Flensing a fin whale (21.7m), Grytviken 1927.

Institute of Oceanographic Sciences

Flensing a southern right whale (14.4m), Grytviken 1931.

Institute of Oceanographic Sciences

A modern whalecatcher, Durban.

Institute of Oceanographic Sciences

Although he was unable to make all the alterations in gear and equipment which he wanted to, Captain Larsen headed for the Ross Sea again the following season. Unfortunately he died before the icefield was penetrated, and in accordance with his wishes he was embalmed and brought home to Norway at the end of the expedition. The latter proved successful, 32,000 barrels of oil being the result. Larsen had shown what could be done in the Ross Sea even though he was restricted by having his factory ship in a safe anchorage. He could have achieved much more if he had been able to operate in the open sea.

The problem of open-sea whaling was being pressed more and more on the whaling companies as the whales disappeared from coastal waters. Longer distances for the catchers meant fewer whales handled, which in turn meant less profit. The factory ship had to become independent of the shore and had to be capable of handling the biggest whales on board. It had to carry all supplies for itself and its catchers and had to be able to carry out any necessary maintenance. If those problems were solved, the most lucrative of all whales, the blue, as yet hardly touched, could be exploited. Whaling in the open sea meant no licences or leases to pay and no regulations to follow.

The first real success at finding a new method came to a Narvik company which constructed, to Captain Peter Sorelle's design, a stern slipway on the SS *Lancing*. The dead whale was hauled up the slipway by a huge, clawlike grab, and completely processed on board. The *Lancing* whaled successfully in the South Atlantic in 1926. Now there was no escape for the whale.

A whalecatcher delivering sperm whales to a slipway, Durban. The dead whales were brought to a slipway close to the harbour entrance, but they had to be removed within a few hours otherwise the authorities had any offending carcase towed out to sea.

Institute of Oceanographic Sciences

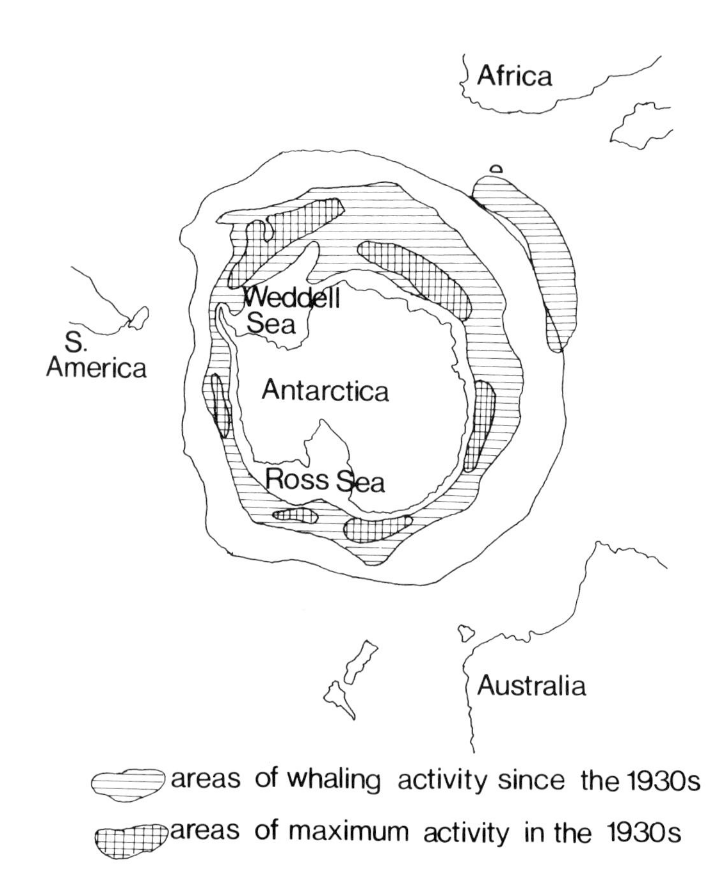

In 1929 the Norwegians started pelagic whaling in the North Atlantic and continued until 1954. Pelagic whaling in the North Pacific started with a Russian expedition in 1936 and, apart from interruption through war, has continued ever since. But it was in the far south that the pelagic whaler made its deepest impression and altered the focal point of the industry. The number of whales killed in the Antarctic topped 10,000 in the 1924–25 season and it never fell below that, except for 1931–32, until the outbreak of World War II. The pelagic whaler brought a big increase in the slaughter, which passed the 40,000 mark in 1930–31, with the peak years between 1934 and 1940 when the number of whales killed never fell below 30,000.

A vast amount of capital was invested in the industry and companies were always on the lookout for improvements to give greater efficiency. The practice of converting the old liners and cargo vessels was abandoned and vessels were specially designed and built for whaling. The first two of these, the *Vikinger* and *Terje Viken*, each between 20,000 and 25,000 tons, arrived in the Antarctic for the 1930–31 season. Pelagic whalers became bigger and reached 30,000 tons in 1937–38 when the Germans sent the *Unitas* to the Antarctic. The *Kosmos*, one of the largest of prewar whalers, carried 300 men. To provide an adequate supply of fresh water, equipment was installed to distil 250 tons per day. The *Kosmos* had half an acre of open deck on which the whale was flensed after being towed up the slipway. She had hold space for 155,000 barrels of oil, and on her first expedition took 119,434 barrels from 1800 whales. She carried an aeroplane for spotting whales, an idea first tried out in 1929.

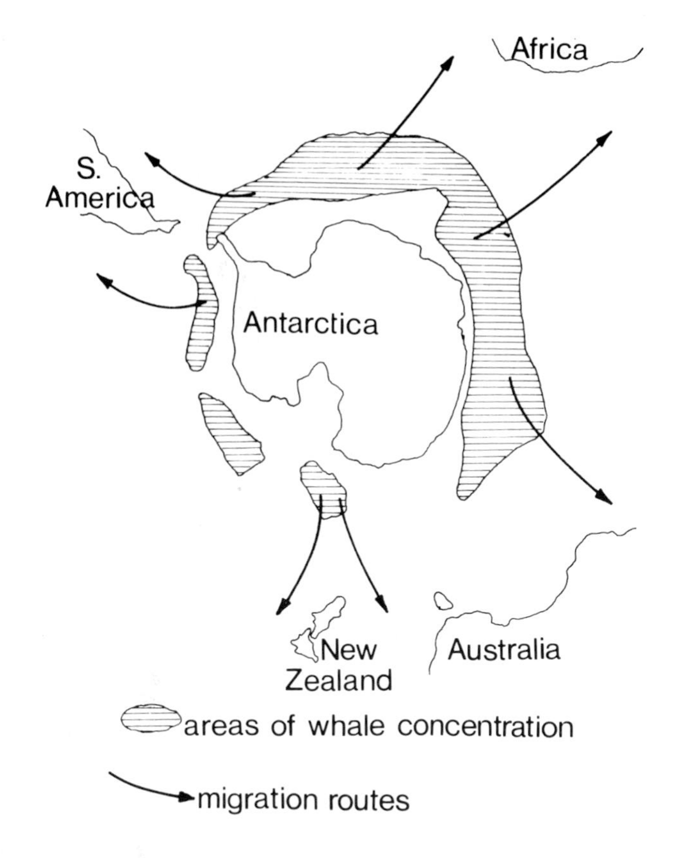

The whalecatchers used between the two wars were small, highly manoeuvrable vessels about 120ft long with a beam of about 23ft and a draught of about 15ft. The accumulators to take the strain of the whale were a series of springs and the lines, made from 6½in manilla, 120 fathoms long, ran via the masthead to a position below deck where they were coiled. The foregoer, made of 4½in Italian hemp, was coiled below the muzzle of the harpoon gun. To counteract the possibility of their breaking under the strain of the whale's struggles, the harpoons, which were 6ft long and weighed 120lb, were made of the best Swedish steel. The 3in bore gun was 45in long and the charge of 14oz of gunpowder gave it a range of 25yds. The phial of sulphuric acid which Foyn had used was replaced by a time-fuse, and muzzle-loaders gave way to breech-loaders in 1925. In 1925–26 quicker access to the gun platform was gained when a Norwegian gunner first introduced a cat-walk leading directly from the bridge to the platform.

With large profits involved it was only natural that the people concerned did not want to listen to the cries for protection of the whale and for a regulation of the industry. Companies and governments turned a deaf ear or were only half-hearted in implementing any suggestions. Nevertheless various bodies kept on with their campaigns for whale protection. It was realised that a deeper knowledge of the whale, its habits, its biological development and its place in the cycle of marine life would give a better foundation on which to base any regulations. In 1923 the British set up the Discovery Committee with the object of promoting and carrying out scientific observations of whales in the Falkland Islands Dependencies. It worked under the guidance of the

Humpback whale, Grytviken.

Institute of Oceanographic Sciences

British Colonial Office in co-operation with the Ministry of Agriculture and Fisheries and the Natural History Department of the British Museum. Funds for its work were raised from taxes imposed on whale oil processed in the dependencies. Its first establishment was a marine biological station in King Edward's Cove opposite the Grytviken whaling station in South Georgia. Captain R F Scott's old ship, *Discovery*, was converted to enable her to carry out marine research, and she sailed in 1925. She returned two years later after a highly successful voyage which persuaded the committee to replace her with a new oil-fired vessel, the Royal Research Ship *Discovery II*. A third ship RRS *William Scoresby*, was added later. The Discovery Committee continued its valuable research until 1949 when its work was taken over and ably continued by the National Institute of Oceanography.

Whale studies by Norwegians began in 1928 and led to the founding of the Norwegian State Institute for Whale Research in 1934. Important statistical information is published annually by the Committee for Whaling Statistics in Oslo which was formed in 1929. In June 1929 the Norwegian Government made it a crime for Norwegian whalers to kill the right whale and any cow accompanied by a calf, and also stipulated that crew should not be paid on the number of whales killed. On 3 April 1930 a committee met in Berlin to consider 'whether and in what terms, for what species and in what areas, international protection of marine fauna could be established', and a list of proposed regulations was drawn up. The first steps towards conservation had been taken, but without implementation they were bound to fail. The whale continued to be slaughtered in such numbers that its survival was in danger.

Fin whale, Grytviken 1926.

Institute of Oceanographic Sciences

In 1937 the first International Convention was signed by Norway, Great Britain, South Africa, the United States, Argentina, Australia, New Zealand, Germany and Ireland. The main points laid down were (i) the opening and closing dates for Antarctic whaling – 8 December and 7 March – and (ii) the complete protection of right whales and grey whales and of others below a certain size. All nations agreed to the regulations but there was no means of enforcing them and the Convention was powerless when one South African and four Japanese expeditions ignored them. The following year the killing of humpbacks by pelagic whalers was forbidden.

When the conference was held in London in 1939 only five nations, the United States, Germany, Japan, Norway and Great Britain, renewed the 1938 agreement, but before long men's minds were occupied by the war which burst upon Europe and finally engulfed the world. Nations pressed their factory ships into transport work and their catchers became escort vessels and minesweepers. At the start of hostilities there were 41 factory ships in the world. At the end of the war only 9 were intact; 28 had been sunk, one of which was a cargo ship and three being tankers. The Japanese whaling fleet, which had expanded rapidly to become the third largest just prior to the war, had been destroyed. Britain had lost all her factory ships; the Norwegians had three left out of thirteen. Germany had three left out of five and these were being made ready by the Allies for a resumption of whaling.

It was obvious from the world shortage of oils and fats that there would be a big attack on the whales as soon as possible. With ideas for regulation to try to preserve whale stocks, representatives of Great Britain, the United States, South Africa, Australia, Canada, New Zealand, Eire and Norway met in London 7 February 1944 and specified that the 1938 regulations should operate when whaling was resumed. They also introduced a new measurement for quotas – the Blue Whale Unit or BWU as it became known.

The BWU was based on the quantity of oil obtained from an average-sized blue whale and the equivalents were fixed as 1 blue = 2 fins = 2½ humpbacks = 6 sei. The number of BWUs for the first season of pelagic whaling after the war was fixed at 16,000. The figure was based on the average prewar catch of 24,000 BWUs. The weakness of this system lay in the fact that limits per species were not specified and therefore overkilling to the point of extermination could happen to a particular species. The BWU never got to the crux of the matter – the preservation of the whale so that stocks could be exploited without extermination – but it was to remain in force until 1972.

Sei whale, Grytviken 1927.

Institute of Oceanographic Sciences

Sei whale on the slipway of a whaling station.

Institute of Oceanographic Sciences

The modern factory ship, like this Japanese vessel, reaches over 30,000 tons. It is designed to deal with the carcase in the shortest possible time and is completely self-sufficient so that it can take the whales in the open sea.

By courtesy of Nippon Suisan Kaisha Ltd Tokyo

10 The Death of the Whale

The war left a world food shortage, especially of edible oils, so preparations to put as many expeditions as possible into the Antarctic for the 1945–46 season were speeded up. Nine factory ships sailed: 6 from Norway, 2 from Britain, and 1 from South Africa. The following season 15 were on the whaling grounds, Norway sending 7, Britain 3, South Africa 1, the Netherlands 1, the USSR 1 and Japan 2. It is significant that defeated Japan was back in the Antarctic so soon after the war. After some discussion and various objections, Japan was given permission to rebuild her whaling fleet in order to combat the food shortage in Japan by supplying whale meat for human consumption. So the great whale hunt was on and man once again overfished. The respite which the whale had enjoyed during the war was soon eliminated and the decrease in whale stocks alarmed the conservationists and those who saw overfishing as the death-knell of the industry.

In December 1946 the International Convention for the Regulation of Whaling was held in Washington. Its purpose was to safeguard the 'natural resources represented by the whale stocks' recognising the fact 'that it is essential to protect all species of whales from further overfishing'. A schedule drawn up forbade the killing of grey whales or right whales, 'calves or suckling whales or female whales which were accompanied by calves or suckling whales', and certain other whales in certain areas. The BWU was written into the schedule. The Convention established the International Whaling Commission which was composed of delegates of the signatory nations and was to meet annually to review the schedule and make any necessary amendments.

The first meeting of the Commission was held in London in 1949, at which a scientific sub-committee, whose findings were to be the basis of future amendments, was established. There were several weaknesses in the schedule, among them the BWU system, but even after it was realised that regulation by species would be more effective, a point recommended by the Scientific Committee from 1963 onwards, it did not come into being until the twenty-fourth meeting of the International

Sectional drawing of a factory ship, 1946.

Illustrated London News

Whaling Commission in 1972. Another weakness of the schedule allowed member nations who did not want to abide by any item or amendment to lodge an objection which automatically gave exemption from that regulation. Objections were continually raised so that much of the work of the Commission was rendered ineffective.

Enforcement of the schedule was the responsibility of individual governments who appointed their own inspectors, but this left a great many loopholes. There was continual pressure from the 1950s for the introduction of an international observer scheme, by which countries exchanged inspectors, but it did not come into operation fully until 1972.

Apart from lacking the authority over its own members the Commission was unable to do anything about whaling by non-member countries. In 1952 Chile, Peru and Ecuador established an independent Permanent Commission for the Exploitation and Conservation of the Marine Resources of the South Pacific. Now there were two whaling groups and they were not following the same rules. This left convenient loopholes such as that exploited by Japan in 1967 when the International Whaling Commission banned the taking of blue whales. Japan allowed her whalers to form companies in Chile and hunt the blue whale under the Chilean flag of convenience. Moreover, the Commission was slow, or failed, to implement fully many of the recomendations of its sub-committees, largely due to rules of its own making. It needed powers of enforcement and means of imposing punishments for illegalities.

But more than anything the main requirement was for all nations to see the folly of overkilling which, as it had done throughout whaling history, could bring many species to the point of extinction. Throughout the years there were to be many protests by numerous bodies to try to make the whaling nations see sense.

Norway was the leading whaling nation after World War II until the 1960s. Britain was second until the end of the 1950s when she was overhauled by Japan. The Russians gradually increased their efforts in the Antarctic and by 1962–63 had moved into second place behind Japan. This was the last season in which the British operated. The Netherlands sent her last expedition the following year and the 1967–68 season saw the last Norwegian expedition except for a small combined floating factory/catcher of 787 gross tons which operated for only one season. The Antarctic whaling grounds were left to the Japanese and Russians.

Whales had been slaughtered in vast numbers. In the Antarctic alone the figure never fell below 30,000 each season from 1947 to 1963. The decline in numbers plus drops in the price of whale products made the cost of whaling uneconomical for the European whalers but the Japanese and Russians, with their more modern equipment, continued to take whales in viable numbers and were helped because of the demand for whale meat for human consumption. Nevertheless the decline in numbers did cause these two countries to start pelagic whaling also in the North Pacific, the Bering Sea and off Kamchatka.

Modern whalecatchers are about 200ft long, displace over 500 tons and are propelled by six-cylinder diesel engines of 2000–3600ihp giving a maximum speed of 18kts. They are highly manoeuvrable vessels, strongly built to stand the buffeting of the most ferocious seas in

A modern Norwegian small whalecatcher.

Author

A model of the whalecatcher *Southern Wheeler* built to a scale of 1/48. This catcher was one of a number built by Smith's Dock Co Ltd, at South Bank, Middlesborough for the South Georgia Co Ltd, and is typical of those developed immediately after the Second World War. Engines developing 1850 hp gave a speed of 16kts. Dimensions were: length 148.5ft, breadth 24.6ft, depth 15.7ft, for a gross tonnage of 427.

Crown Copyright, Science Museum, London

the world and to combat ice conditions in the lonely Polar regions. They are crewed by twelve men, with cramped but comfortable quarters. Ranging as much as a hundred miles, they are constantly in touch with the factory ship by mens of powerful radio transmitters and receivers.

Once the cry of 'She blows, there she blows' comes from the crow's nest at the masthead, the harpooner hurries along the catwalk from the bridge to the gun platform on the bow. Glancing back, he gets the direction of the whale from the man in the crow's nest; from then on, the harpooner demands instant obedience as he controls the speed of his ship by means of the signalling apparatus located on the platform, and the direction by arm signals to the helmsman. The heaving bow sends spray showering over him as the catcher plunges into the turbulent sea; the biting wind tears at him in his exposed position but he steels himself against the forces of nature, concentrating his efforts on hunting the whale.

The whale has little chance against the harpoon gun once the catcher is in range. The gun is perfectly balanced and easily pivoted. Once it is accurately aimed and fired it sends lethal destruction curving over the water towards the victim. When the harpoon has struck, the man on the windlass 'plays' the whale so that the line does not break, and any attempt to run is counteracted by the catcher itself.

The dead whale is pulled alongside the catcher with its tail at the bows; a hollow lance is plunged into the body and compressed air is pumped in so that the whale will float. The tail is heaved from the sea and a steel ring is slipped over it and secured. A small transmitter and a marker, with the catcher's number, are stuck into the body and the whale is cast adrift, to be located by the buoy boat which will fasten a line to the steel ring and tow the whale to the factory ship.

A Japanese whalecatcher in action in Antarctica.

By courtesy of Nippon Suisan Kaisha Ltd, Tokyo

A whalecatcher with a harpooned sei whale under its bow, off Durban.

Institute of Oceanographic Sciences

A catcher bringing whales to the factory ship.

By courtesy of Nippon Suisan Kaisha Ltd, Tokyo

When the buoy boat reaches the factory ship, the whale is put with those from other catchers to wait its turn on the flensing deck. The whale is manoeuvred into the opening in the stern of the factory ship and a huge grab is fastened to its tail by which it is dragged up the slipway. The friction from the huge body is so great that hoses have to be constantly played on to it as it is pulled towards the deck. The flensers are upon it before it has stopped. They make their cuts quickly and skilfully. Hooks are fastened into the skin which the winches peel off; the flensers then tackle the blubber, cutting the body in a special way so that the winches can pull the lumps away as the flensers cut them. The whole carcase is dealt with speedily and systematically, the various parts going down the appropriate holes in the deck to the boilers where the oil and other valuable commodities are extracted. The baleen plates from the upper jaw are the only parts of the whale to be discarded – a far cry from the late eighteenth century when this was the most valuable part.

The deck resembles a clattering chaos as blades flash, winches roar, blood spurts and the deck becomes a slimy, greasy surface, in what is in fact a highly organised operation. No sooner is one whale cut up than the next is being hauled up the slipway to the flensers by the winch-man, whose concentration and skill are essential for everyone's safety, in a highly dangerous occupation. One slip and a man could be severely injured, lose a limb or even his life in the stench which follows the first sweep of the flensing knives.

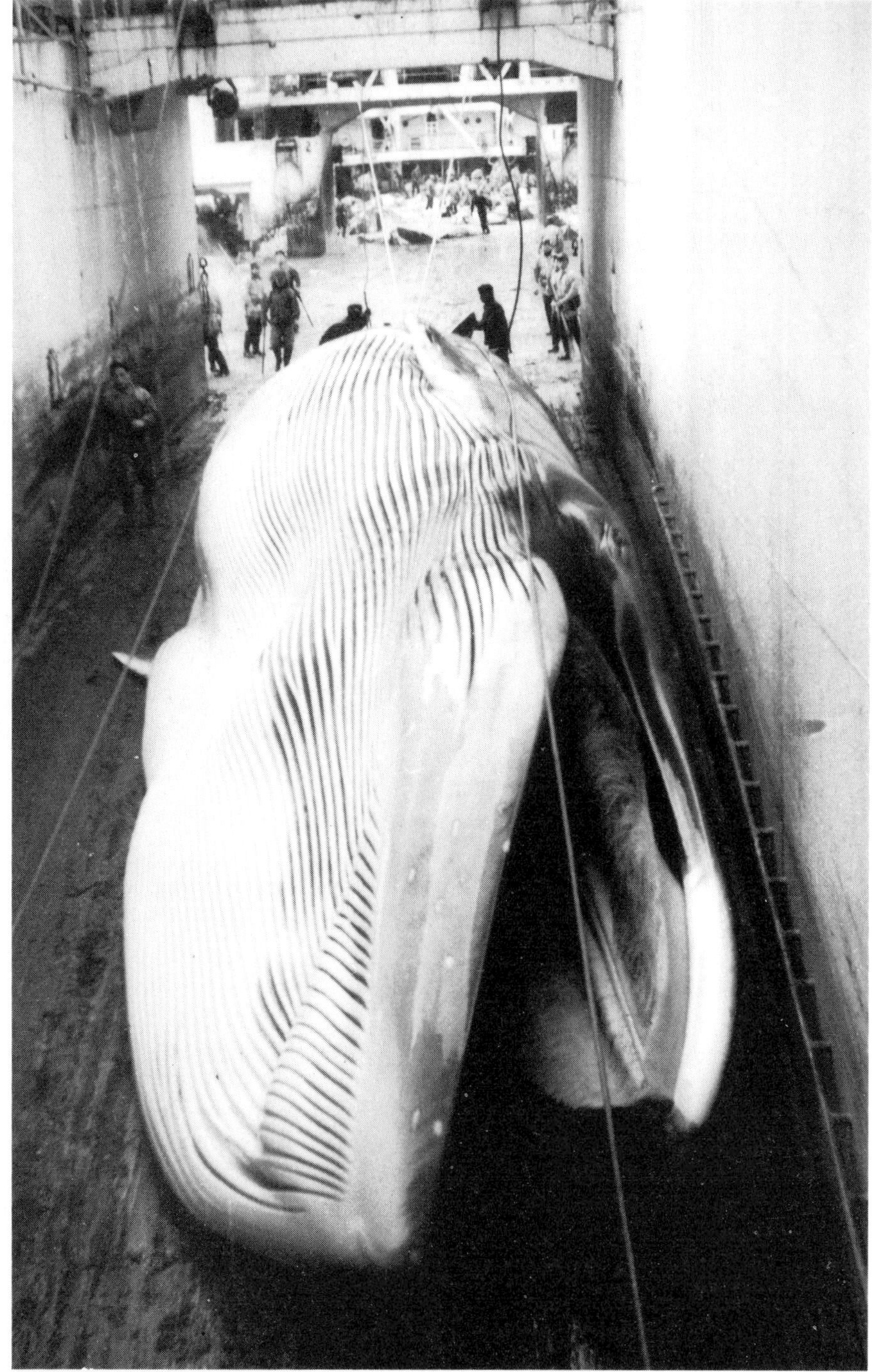

Hauling a whale up the factory ship's slipway. A slipway became necessary so that the largest whales could be taken on to the deck for easier flensing. The stern slipway was designed by Captain Peter Sorelle and first used on the SS *Lancing*. It was operated successfully, though there were teething troubles, and by the 1929/30 season the slipway had become standard equipment. At first men had to fix wire ropes to the dead whale at the bottom of the slipway. This hazardous operation was eliminated with the invention of Gjelstad's whale claw.

By courtesy of Nippon Suisan Kaisha Ltd, Tokyo

Cutting up whales, Iceland.

Author

The process goes on night and day as long as there are whales to be dealt with. Twelve a day is a good average though one whale an hour has been achieved. The ships which deal with whales on this scale are monsters of 20,000 to 30,000 tons with every comfort for their crews of 300 to 400 men. They are fitted with every device – powerful radios, automatic pilots, radar, supersonic sounding apparatus, direction finders, in fact everything and anything which will help them hunt the whale more efficiently as they aim for higher profits.

Apart from the pelagic whaler a few shore stations are still operational. Typical of these is the one at Hvalfjordur, not far from Reykjavik, the capital of Iceland. Started in 1948, it has been profitable ever since. In addition to observing international restrictions certain self-imposed limits were set and have resulted in a thriving industry employing about 100 people.

But not all whalers feel themselves bound by restrictions, and killing still goes on at a rate detrimental to the preservation of stocks. Certain small whaling activities with their own peculiarities still exist in various parts of the world.

Whale flensing, Iceland.

Author

A flenser, Iceland

Author

Flensing, Iceland.

Author

A caeing whale hunt, Torshavn, the Faeroes. The whales are driven into the shallow confined waters of the harbour where the slaughter takes place. The hunt has much tradition and follows certain rituals and patterns which are part of the folklore of the Faeroes. See also overleaf.

By courtesy of Mr Kenneth Williamson

Rowing boats are still used to hunt the sperm whale from the Azores. The Portuguese people of these islands learned about whaling from the Americans, who first exploited the sperm whales in the area in 1765. Offshore whaling in the Azores began about 1832 from the island of Fayal and spread to all the others except Corvo which is almost inaccessible. Their whaleboats are similar to those formerly used by the Americans but, at 38ft, are 10ft longer. This extra length allows for an additional crew member, so there are six oars compared with the American five, two of 16ft, two of 17ft and two of 18ft, the longest being used amidships.

There is a system of lookout points around the coasts, especially built to withstand the wild Atlantic winds. From these positions high on the cliffs, lookouts scan the sea through powerful binoculars. When a whale is sighted, a rocket is fired and the cry of 'Baleia' echoes through the whaleboat stations. Motor launches meet up with the boats, take them in tow and are guided by the lookouts over the radio telephone towards the whales. The launches shut off their engines at a suitable distance from them and the whaleboats move quietly forward using sail, paddles or oars. The hunters follow the old American methods of sperm whaling, except that the harpooner also wields the lance, instead of changing places for the boat-header to make the kill.

At one time the dead whale was dealt with by the whaling crews at try-works erected near the beach, as was the practice in New England in the early eighteenth century, but since the 1950s the whale has been processed in a modern factory equipped with pressure cookers, by workers who do not take part in the hunt.

The blackfish, or caeing whale, is still a source of supply to the people of the Faeroe Islands – 1050 were taken in 1973 – but it is no longer as vital as it once was when a good kill meant ample food for the long winter. Nevertheless the hunt still goes on, following rituals which can be regarded as part of the folklore of the islands.

Once a school of whales (which can vary from a few to hundreds) has been seen, a prearranged signal is given, and the cry sweeps through the islands. At one time there was a system of beacons, of white sheets laid out on hillsides and of runners, conveying the nature and size of the school and the most likely place for a kill. Nowadays the message is relayed by telephone and everyone connected with the hunt leaves his work and takes to the boats or goes to his place on the quayside or beach. Hardly anyone on the island will miss the kill for this is an event traditionally celebrated.

There is a captain in charge of the entire operation who must take into account the state of the tide, the direction of the wind, the amount of daylight left and so on. If a sighting is made late in the day, the whales are driven into a fjord where they are prevented, by the boats, from escaping to sea and where they remain fairly subdued. The captain must decide where the school is to be driven for the kill. The boats form an arc and herd the whales in the required direction. The ideal destination is a beach or the head of some harbour, so that, once stranded, the whales can be easily killed with lances and knives.

Once the terrible slaughter is over, dancing and feasting begin, with many of the islanders wearing national costume. The proceeds from the whales are distributed among the people who use nearly everything as food. Some of it is eaten fresh, whilst the remainder is dried out on special racks or salted, though in some places deep-freezing is now the method of preserving and storing. So today, age-old traditional methods are practiced alongside pelagic ships and up-to-date catchers, perpetuating the long colourful history of whaling, yet at the same time hastening its end.

The fear that the whale might be exterminated caused grave concern and many conservationist bodies throughout the world constantly raised their voices against the continual slaughter and called the International Whaling Commission to task for allowing it to continue by its inability to act. Constant campaigning by numerous organisations and individuals throughout the world brought the plight of the whale sharply before world governing bodies and there grew a genuine feeling and concern for the diminishing stocks of whales, even by many of the Commissioners serving on the International Whaling Commission.

A sperm whale off Durban.

Institute of Oceanographic Sciences

Nations closed down their whaling enterprises or banned the import of whale products as they came to realise that whale hunting had become uneconomic for them and that with substitute products available there was no need to kill the whale for what it could supply.

A resolution, calling for a ten-year world-wide moratorium on all whaling, put forward by the US delegation at the United Nations Conference on the Human Environment in 1972, was carried overwhelmingly, and it was thought that in the face of public opinion the International Whaling Commission would have to act likewise. It failed to do so but did take an important step by abolishing the BWU and establishing quotas. The Scientific Committee believed that with a moratorium on certain species and controlled quotas on others there was no need to introduce a moratorium on all whaling. After this meeting in 1972 there was a moratorium on right whales, blue, grey and humpback whales and this list was widened in 1975 to include other whales in certain areas.

The continual protest at whale killing ws kept up. The Greenpeace Foundation, which had come into being as a result of protestations against nuclear tests, turned its attention to the plight of the whale. Apart from verbal protests the members of Greenpeace also made yearly, active, physical protestations by sailing rubber boats between the whalecatchers and the whales they were pursuing. Their activities, coupled with those of other conservation bodies, such as Friends of the Earth, the Fauna Preservation Society and the World Wildlife Fund, have brought continual pressure to bear on the whaling nations and on the International Whaling Commission.

The 1979 meeting of the Commission brought forward the proposals that a whale sanctuary be formed in the Indian Ocean, and that factory ships be banned from taking the sperm whale – further steps along the way to a complete moratorium which, some say, is the only way to save the whale. In the meantime whaling continues in some form or other, but as whales become harder to find the whaling nations, and rogue whalers operating outside any authority, make less profit than they would like. They may find that whaling becomes totally uneconomic. It is to be hoped that by the time that occurs, or a complete moratorium is imposed, whales have not been slaughtered beyond recovery.

If that happens we will have lost a creature of high intelligence which has fascinated man for centuries.

Sperm whale, Durban.

Institute of Oceanographic Sciences

Bibliography

LOGBOOKS AND JOURNALS

Whitby Museum
Seven Logbooks concerning the Arctic Voyages of Captain William Scoresby Senior of Whitby, England (Facsimile by Explorers Club, New York, 1916–17). Logbooks and journals kept by William Scoresby on nine voyages in the *Henrietta*, on seven of which (1791–97) he was captain, and voyages in the *Dundee* (1798 and 1801), the *Mars* (1817) and the *Fame* (1820 and 1822).
Original Logbooks of Captain William Scoresby Senior's Eight Voyages in the Resolution, 1803 to 1810 inclusive One volume containing journals of 1807–9, copied from originals and examined, corrected and signed by Wm Scoresby Jr; one journal of 1810, copied from original, examined, corrected and signed by Wm Scoresby Jr.
Original Logbooks of Captain William Scoresby Junior's Voyages in the Resolution, 1811 and 1812.
Original Logbooks of Captain William Scoresby Junior's Voyages in the Esk, 1813 and 1814.
Original Logbook of Captain William Scoresby Junior's Voyage in the Baffin 1821.

Hull Public Library
Original Logbooks
1812 *Margaret:* James Hewitt, Commander
1813 *Margaret:* James Hewitt, Commander
1814 *Margaret:* Robert Gascoin, Commander
1815 *Margaret:* Robert Gascoin, Commander
1816 *Margaret:* Robert Gascoin, Commander
1819 *Neptune:* M Munroe, Master
1820 *Neptune:* M Munroe, Master
1821 *Neptune:* M Munroe, Master
1823 *Neptune:* M Munroe, Master
1824 *Exmouth:* Matthew Wright, Master
1825 *Brunswick:* William Blyth, Captain
1826 *Brunswick:* William Blyth, Captain
1826 *Cumbrian:* M Munroe, Master
1827 *Brunswick:* William Blyth, Captain
1827 *Cumbrian:* M Munroe, Master
1827 *Laurel:* William Manger, Captain
1828 *Laurel:* William Manger, Captain
1828 *Progress:* Edward Dannatt, Commander
1828 *William:* T North, Captain
1829 *Brunswick:* William Blyth, Captain
1830 *Eagle:* Matthew Wright, Master
1831 *Andrew Marvel:* Matthew Wright, Master
1852 *Orion:* Emmanuel Wells, Captain

Hull Maritime Museum
Original Logbooks
1821 *Royal George:* Rickett, Captain
1822 *Duncombe:* Corbett, Captain
1831 *Dordan:* Edward Willis, Master
1832 *Volunteer:* Henry Parish, Commander
1833 *Volunteer:* Henry Parish, Commander
1834 *Volunteer:* Henry Parish, Commander
1857 *Swan:* Joseph Taylor, Commander
1859 *Truelove:* William Wells, Commander
1860 *Truelove:* John Parker, Commander
1861 *Truelove:* William Barron, Commander
1871 *Erik:* J B Walker, Commander
Minute Book Owners of ships belonging to the Port of Hull employed in the Greenland and Davis Strait Fisheries 1813
Papers 1803–1808 of Caerwent Whaler and Privateer
Journals of Charles Edward Smith of the Diana 22 March 1866–26 April 1867. These were later edited by his son and published in 1922 under the title *From the Deep of the Sea*
Copy of MS kept by W Eden Cass, a Surgeon on board the Brunswick 1824

Hull Trinity House
Original Logbooks
1812–15 *Comet:* Abel Scurr, Captain
1828 *Dordan:* William Linskill, Captain
1828 *Ariel:* Richard Rogers, Captain
1838 *Andrew Marvel:* George Silcock, Captain

BOOKS

Aldrich, Herbert L: *Arctic Alaska and Siberia* (Chicago, 1889)
Altonaer Museum in Hamburg: *Wale und Walfang* (Hamburg, 1975)
Anderson, Florence Bennett: *Through the Hawse-Hole* (New York, 1932)
Andrews, Roy Chapman: *Whale Hunting with Gun and Camera* (New York, 1916)
Andrews, Roy Chapman: *Ends of the Earth* (New York, 1929)
Aristotle: *Historia Animalium.* Translated by D'Arcy W Thompson (Oxford, 1910)
Armstrong, Warren: *The True Book about Whaling*
Ash, Christopher: *The Whaler's Eye* (New York, 1962)
Ashley, Clifford W: *The Yankee Whaler* (1938)
Baird, Patrick D: *The Polar World* (1964)
Barron, W: *Old Whaling Days* (Hull, 1895; reprinted 1970)
Beal, T: *The Natural History of the Sperm Whale* (1839; reprinted 1973)
Beal, T: *A Few Observations on the Natural History of the Sperm Whale* (1835; reprinted 1976)
Beddard, F E: *A Book of Whales* (1900)
Bennett, A G: *Whaling in the Antarctic* (New York, 1932)
Bennett, Frederick D: *Narrative of a Whaling Voyage Round the Globe from the Year 1833 to 1836* (1840)
Blainey, Geoffrey: *The Tyranny of Distance* (1968)
Blond, Georges: *The Great Whale Game* (1954)
Bootes, Henry H: *Deep Sea Bubbles* (New York, 1929)
Brewington, M V and Dorothy: *Kendal Whaling Museum Paintings* (Sharon, Massachusetts, 1965); *Kendal Whaling Museum Prints* (Sharon, Massachusetts, 1969)
Brown, R N R: *Spitsbergen* (1920)
Browne, J Ross: *Etchings of a Whaling Cruise* (New York, 1846; reprinted Cambridge, Massachusetts, 1968)
Budker, P: *Whales and Whaling* (1958)
Bull, W J: *Cruise of the Antarctic to the South Polar Regions* (1896)
Bullen, Frank T: *The Cruise of the Cachalot* (1899)
Burrows, Fredrika Alexander: *The Yankee Scrimshanders* (Taunton, Massachusetts, 1973)
Burton, Robert: *The Life and Death of Whales* (1973)
Calendar of State Papers
Campbell, John: *A Narrative of the Shipwreck of the Shannon of Hull on her passage to Davis Strait 26 April 1832* (Edinburgh, 1833)

Cameron, Ian: *Antarctica The Last Continent* (1974)
Cary, M and Warmington, E H: *The Ancient Explorers* (1929)
Chase, Owen: *Shipwreck of the Whaleship Essex* (Gloucester, Massachusetts, 1972)
Chatterton, E K: *Whalers and Whaling* (1925)
Cheever, H T (edited by W Scoresby): *The Whaleman's Adventures* (1850)
Chippendale, Harry Allen: *Sails and Whales* (Cambridge, Massachusetts, 1951)
Churchill, J and A: *Collection of Voyages* (1744)
Ciriquiain-Gaiztarro, M: *Los Vascos en la Pesca de La Ballena* (San Sebastian, 1961)
Clarke, Robert: *Open Boat Whaling in the Azores Discovery Reports,* xxvi, 281–354 (1954)
Cockrill, W Ross: *Antarctic Hazard* (1955)
Coffey, D J: *The Encyclopedia of Sea Mammals* (1977)
Colnett, J A: *Voyage Round Cape Horn to the Pacific Ocean* (1798; reprinted Amsterdam, 1968)
Colwell, Max: *Whaling around Australia* (1970)
Committee for Whaling Statistics: *International Whaling Statistics* (Various; Oslo, annually)
Conway, Sir W Martin: *No Man's Land* (1906)
Conway, Sir W Martin (editor): *Early Dutch and English Voyages to Spitsbergen in the 17th Century* (1904)
Cook, John A: *Pursuing the Whale* (1926)
Cousteau, Jacques-Yves: *The Whale* (1972)
Cowlin, Dorothy: *Greenland Seas* (Leeds, 1965)
Crevecoeur, Jean Hector St John de: *Letters from an American Farmer* (1782)
Crisp, Frank: *The Adventure of Whaling* (1954)
Dakin, W J: *Whalemen Adventurers* (Sydney, 1934)
Dautert, Erich: *Big Game in Antarctica* (Bristol, 1937)
Davis, William M: *Nimrod of the Sea* (1874; reprinted North Quincy, Massachusetts, 1972)
Dow, George Francis: *Whale Ships and Whaling* (Salem, 1925; reprinted New York, 1967)
Dulles, Foster Rhea: *Lowered Boats* (1934); *Harpoon* (1935)
Dumas, Alexandre and Maynard F: *The Whalers* (1937)
Earle, Walter K: *Scrimshaw* (Cold Spring Harbor, New York, 1957)
Edwards, Everett J and Rattray, Jeannette Edwards: *Whale Off!* (New York, 1932)
Elking, H: *A View of the Greenland Trade Whale-Fishery with the National and Private Advantage Thereof* (1722)
Ely, Ben-Ezra Stiles: *There She Blows* (Philadelphia, 1849; reprinted Middletown, Connecticut, 1971)
Fanning, E: *Voyages Round the World* (New York, 1833)
Ferguson, Henry: *Harpoon* (1932)
Flayderman, E Norman: *Scrimshaw and Scrimshanders* (New Milford, Connecticut, 1972)
Forbes, Alan: *Whale Ships and Whaling Scenes as Portrayed by Benjamin Russell* (Boston, Massachusetts, 1955)
Fraser, F C: *British Whales, Dolphins and Porpoise* (1966)
Gad, Finn: *History of Greenland* Vol. I (1970) and Vol II (1973)
Gardiner, Alice Cushing and Osborne, Nancy Cabot: *Father's Gone A-Whaling* (New York, 1926)
Gardner, Erle Stanley: *Hunting the Desert Whale* (1963)
Gardner, Will: *Three Bricks and Three Brothers* (Cambridge, Massachusetts, 1945)
Gaskin, D E: *Whales, Dolphins and Seals* (1972)
Gaskin, Robert Tate: *The Old Seaport of Whitby* (Whitby, 1909)
Gerstaecker, F: *Journey Round the World* 1847–52 (1853)
Gilbreth, Frank B: *Of Whales and Women* (1957)
Grant, Gordon: *Greasy Luck* (1932)
Gray: *The Manner of Whale Fishing in Greenland.* Given in by Mr Gray to Mr Oldenbury for the Society (1663), Sloane MS698, ff346–366, British Museum
Gray, D and J: *Report on New Whaling Grounds in the Southern Seas* (Aberdeen, 1874)
Gray, R W: *Peterhead and the Greenland Sea* (Aberdeen, 1942)
Grierson, John: *Air Whaler* (1949)
Haig-Brown, Roderick: *The Whale People* (1962)
Hakluyt, R: *The Principal Navigations, Voyages, Traffiques and Discoveries of the English Nation* (1599)
Haley, Nelson Cole: *Whale Hunt* (1950)
Halsey, Helen: *Incident on the Bark Columbia* (Cummington, Massachusetts, 1941)
Hammond, Walter: *Mutiny on the Pedro Varela* (Mystic, Connecticut, 1956)
Hare, Lloyd C M: *Salted Tories* (Mystic, Connecticut, 1960)
Hardy, Sir A C: *Great Waters* (1967)
Harmer, S F: *History of Whaling,* Proc Linn Soc, London, Session 140, 51–95 (1928); *Southern Whaling* Proc Linn Soc, London, Session 142, 85–163 (1931)
Hawes, Charles Boardman: *Whaling* (1924)
Hegarty, Reginald B: *Returns of Sailing Vessels Sailing from American Ports* 1876–1928 (New Bedford, 1959)
Henderson, David S: *Fishing for the Whale* (Dundee, 1972)
Hinton, M A C: *Report on the Papers Left by the Late Major Barrett-Hamilton, Relating to the Whales of South Georgia* 57–209 (1925)
Hohman, Elmo Paul: *The American Whaleman* (New York, 1928; reprinted New York, 1970)
Hopkins, William John: *She Blows! And Sparm at That!* (Boston, 1922)
Housby, Trevor: *The Hand of God* (1971)
Howland, Chester S: *Thar She Blows* (New York, 1951); *Whale Hunters Aboard the Grey Gold* (Caldwell, Idaho, 1957)
Hoyt, Edwin P: *The Mutiny on the Globe* (New York, 1975)
Hyde, Michael: *Arctic Whaling Adventures* (1955)
International Whaling Commission: *Schedule to the International Whaling Convention 1946, and Various Amendments*
Jackson, Gordon: *The British Whaling Trade* (1978)
Jenkins, J T: *A History of the Whale Fisheries* (1921); *Whales and Modern Whaling* (1932); *Bibliography of Whaling* (1948)
King, H G R: *The Antarctic* (1969)
Kirk, R with Daugherty, Richard D: *Hunters of the Whale* (New York, 1974)
Laing, Alexander: *American Ships* (New York, 1971)
Laing, John: *An Account of a Voyage to Spitsbergen* (1815)
Lamont, James: *Seasons with the Sea Horses* (New York, 1861)
Laubenstein, William J: *The Emerald Whaler* (Indianapolis, 1960)
Lawson, Will: *Blue Gum Clippers and Whale Ships of Tasmania* (Melbourne, Australia, 1949)
Leavitt, John F: *The Charles W Morgan* (Mystic, Connecticut, 1973)
Lilley, Harry R: *The Path Through Penguin City* (1955)
Lipton, Barbara: *Whaling Days in New Jersey* (Newark, 1975)
Liversidge, Douglas: *The Whale Killers* (1963)
Lubbock, Basil: *Arctic Whalers* (Glasgow, 1937)
Mackintosh, N A: *The Stocks of Whales* (1965)
McLaughlin, W R D: *Call to the South* (1962)
McNab, Robert: *The Old Whaling Days* (Melbourne, 1913)
Markham, Albert Hastings: *A Whaling Cruise to Baffin Bay* (1874)

Markham, C R: *The Threshold of the Unknown Region* (1876); *On the Whale-Fishery of the Basque Provinces of Spain* (1881); The Voyages of William Baffin 1612–1622 (1881)
Martin, Kenneth R: *Delaware Goes Whaling 1833–1845* (Greenville, Delaware, 1974); *Whalemen and Whaleships of Maine* (Brunswick, Maine, 1975)
Matthews, L H: *South Georgia* (Bristol 1931); *The Whale* (1968); *The Natural History of the Whale* (1978)
Melville, Herman: *Moby Dick* (New York, 1851; Oxford University Press 1920; reprinted 1957)
Mielche, Hakon: *There She Blows!*
Moffett, Robert K and Martha L: *The Whale in Fact and Fiction* (1967)
Morey, George: *The North Sea* (1968)
Morison, Samuel Eliot: *The Maritime History of Massachusetts* (Boston, 1941)
Morley, F V and Hodgson, J S: *Whaling North and South* (1927)
Morris, Desmond: *The Mammals* (1965)
Mowat, Farley: *A Whale for the Killing* (1973)
Murdoch, W G B: *From Edinburgh to the Antarctic* (1894); *Modern Whaling and Bear Hunting* (1917)
Murphy, R C: *Logbook for Grace* (New York, 1947)
Naess, Øyvind: *Hvalfangerselskapet Globus AS 1925–1950* (Narvik, 1951)
Nayman, Jacqueline: *Whales, Dolphins and Man* (1973)
Norman, J R and Fraser, F C: *Giant Fishes, Whales and Dolphins* (1937)
Olmsted, Francis Allyn: *Incidents of a Whaling Voyage* (New York, 1841; reprinted New York, 1969)
O'May, Harry: *Wooden Hookers of Hobart Town and Whalers out of Van Diemen's Land* (Tasmania, nd)
Ommanney, F D: *South Latitude* (1938); *Lost Leviathan* (1971)
Paulding, Lt Hiram: *Journal of a Cruise of the United States Schooner Dolphin in Pursuit of the Mutineers of the Whale Ship Globe* (New York, 1831; reprinted London, 1970)
Philip, J E: *Whaling Ways of Hobart Town* (Hobart, Tasmania, 1936)
Pine, W: *History and Antiquities of the City of Bristol* (Bristol, 1789)
Pliny: *The Naturall Historie of C Linius Secundus*
Preston, C: *Captain William Scoresby* (Whitby, 1964)
Purchas, S: *His Pilgrims* (1613–25)
Purrington, Philip F: *4 Years A-Whaling* (Barae, Massachusetts, 1972)
Ricketson, Annie Holmes: *Mrs Ricketson's Whaling Journal* (New Bedford, 1958)
Riggs, Dionis Coffin: *From Off Island* (New York, 1940)
Robertson, R B: *Of Whales and Men* (New York, 1954)
Robinson, F K: *History of Whitby* (1860)
Robotti, Frances Diana: *Whaling and Old Salem* (New York, 1962)
Ruhen, Olaf: *Harpoon in my Hand* (Sydney, 1966)
Sanderson, John: *A Voyage from Hull to Greenland* (Kingston-upon-Hull, 1789)
Sanderson, Ivan T: *Follow the Whale* (1958)
Sawtell, Clement Cleveland: *The Ship Ann Alexander of New Bedford* (Mystic, Connecticut, 1962)
Scammon, Charles M: *The Marine Mammals of the Northwestern Coast of North America* (San Francisco, California, 1874; reprinted New York, 1968)
Scheffer, Victor B: *The Year of the Whale* (1970)
Schevill, William E: *The Whale Problem* (Cambridge, Massachusetts, 1974)
Schmitt, Fredrick P: *Mark Well the Whale!* (Port Washington, NY, 1971); *The Whale's Tale* (Chippenham, Wilts, 1975)
Scoresby, William Jr: *An Account of the Arctic Regions with a History and Description of the Northern Whale-Fishery* (1820; reprinted Newton Abbot, 1969); *Journal of a Voyage to the Northern Whale-Fishery 1822 in Baffin of Liverpool* (Edinburgh, 1823); *Loss of the Esk and Lively* (Whitby, 1826); *My Father* (1851)
Scoresby-Jackson, R E: *The Life of William Scoresby* (1861)
Semmes, Admiral Raphael: *Service Afloat* 1887)
Shaw, Jeffrey: *Whitby Lore and Legend* (Whitby, 1952)
Sherman, Stuart C: *The Voice of the Whaleman* (Providence, 1965)
Slijper, E J: *Whales* (1962)
Smith, Charles Edward: *From the Deep of the Sea* (1922)
Solyanik, A: *Cruising in the Antarctic* (Moscow, 1956)
Southern, H N: *The Handbook of British Mammals* (1964)
Spears, John R: *The Story of the New England Whalers* (New York, 1908)
Stackpole, Edouard: *Mutiny at Midnight* (1944); *The Sea Hunters* (Philadelphia, 1953); *Whales and Destiny* (University of Massachussetts, 1972)
Stamp, Tom and Cordelia: *William Scoresby* (Whitby, 1976)
Starbuck, Alexander: *History of the American Whale Fishery* (Washington, DC, 1878; reprinted New York, 1964)
Tilton, George Fred: *Cap'n George Fred Himself* (New York, 1929)
Tomlin, A G: *Mammals of the USSR and Adjacent Countries – Vol IX: Cetacea* (Moscow, 1957, Jerusalem 1967)
Turner, Sir W: *The Marine Mammals in the Anatomical Museum of the University of Edinburgh* (1912)
Vamplew, Wray: *Salvesen of Leith* (Edinburgh, 1975)
Veer, Gerrit De: *The Three Voyages of William Barents to the Arctic Regions 1594, 1595 and 1596* (1853)
Venables, Bernard: *Baleia* (1968)
Verrill, A Hyatt: *The Real Story of the Whaler* (New York, 1916)
Villiers, A J: *Whaling in the Frozen South* (Indianapolis, 1925); *Vanished Fleets* (1931); *Whalers of the Midnight Sun* (1934)
Walker, Ernest P: *Mammals of the World* (Baltimore, 1964)
Warinner, Emily V: *Voyage to Destiny* (Indianapolis, 1956)
Watson, Arthur C: *The Long Harpoon* (New Bedford, 1929)
Weatherall, Richard: *The Ancient Port of Whitby and its Shipping* (Whitby, 1908)
Wells, John C: *The Gateway to the Polynia* (1873)
Wheeler, James Cooper: *There She Blows* (New York, 1909)
Whipple, A B C: *Yankee Whalers in the South Seas* (1954)
White, Adam: *A Collection of Documents on Spitsbergen and Greenland* (1855)
Wilkinson, David: *Whaling in Many Seas* (1905)
Williams, H: *One Whaling Family* (1964)
Williamson, Kenneth: *The Atlantic Islands* (1948)
Winterhoff, Edmund: *Walfang in der Antarktig* (Oldenburg, 1974)
Wise, Terence: *To Catch a Whale* (1970)
Young, Rev George: *A Picture of Whitby* (Whitby, 1840)
Zenkovich, B A: *Whales and Whaling* (Moscow, 1952)
Zorgdrager, C G: *Alte und neue Grönländische Fischerei und Wallfischfang* (1st German edition Leipzig, 1723; reprinted Kassel, 1975)

Index